Concepts and techniques in genomics and proteomics

Published by Woodhead Publishing Limited

Woodhead Publishing Series in Biomedicine

Published by Woodhead Publishing Limited

Published by Woodhead Publishing Limited

Published by Woodhead Publishing Limited

Woodhead Publishing Series in Biomedicine: Number 5

Concepts and techniques in genomics and proteomics

Nachimuthu Saraswathy and
Ponnusamy Ramalingam

WOODHEAD
PUBLISHING

Oxford   Cambridge   Philadelphia   New Delhi

Published by Woodhead Publishing Limited

Woodhead Publishing Limited, 80 High Street, Sawston, Cambridge, CB22 3HJ, UK
www.woodheadpublishing.com
www.woodheadpublishingonline.com

Woodhead Publishing, 1518 Walnut Street, Suite 1100, Philadelphia, PA 19102-3406, USA

Woodhead Publishing India Private Limited, G-2, Vardaan House, 7/28 Ansari Road,
Daryaganj, New Delhi – 110002, India
www.woodheadpublishingindia.com

First published in 2011 by Biohealthcare Publishing (Oxford) Limited; republished in 2012 by
Woodhead Publishing Limited
ISBN: 978-0-08-101730-2 (print); ISBN 978-1-908818-05-8 (online)
Woodhead Publishing Series in Biomedicine ISSN 2050-0289 (print); ISSN 2050-2097 (online)

Typeset by RefineCatch Limited, Bungay, Suffolk
Printed in the UK and USA

Published by Woodhead Publishing Limited

Contents

Published by Woodhead Publishing Limited

Published by Woodhead Publishing Limited

Published by Woodhead Publishing Limited

Published by Woodhead Publishing Limited

Published by Woodhead Publishing Limited

List of figures

Published by Woodhead Publishing Limited

Published by Woodhead Publishing Limited

List of tables

Preface

Genes and proteins are the storehouse of information and executers of cellular life processes, respectively. Genome sequencing revealed the blueprint of life of many organisms. 'Omics' has become an order of the day. Information derived from genome sequencing projects is exploited to understand the living organisms in a better way. Knowledge from these fields will be helpful in the research and manufacture of new drugs and diagnostic methods.

Genomics and proteomics are newer fields in modern biology which help us to understand the living organisms as a whole. These two fields were developed based on the concepts that existed before but now they have been applied to high throughput techniques. Genomics is mainly concerned with the organization of genes and genomes, the mapping of genomes, genome sequencing, and the annotation of genomes. Proteomics deals with proteins expressed in a cell at different times, post-translational modifications, protein–protein interactions, etc.

This book covers the important techniques that are used in genomics and proteomics. Investigation of these latest techniques and their applications in various fields will help undergraduate students enhance their current perception on biology. The contents of this book have been developed from our teaching to undergraduate classes on genomics and proteomics for the past six years. This book is designed for undergraduate and postgraduate students of life sciences and biotechnology.

The aim of this book is to introduce the concepts of genomics and proteomics and high throughput techniques used in these fields. We are confident that the book will serve the purpose of being useful to students. Every effort has been taken to explain each technique in a simple manner, and flow charts are also provided.

The contents of this book are designed for the beginners in this field. The information obtained from genome projects are being used in many

Published by Woodhead Publishing Limited

field such as pharma, medicine, biomedical, agriculture, etc. It is important to have books like this to explain the concepts and techniques to various academicians who need additional knowledge. Even experienced scientists in this field can update their knowledge with the modern concepts.

Published by Woodhead Publishing Limited

Acknowledgements

We must thank a number of people who are involved in genomics and the proteomics area of research throughout the world. Without their published results this book would not have been possible. First, we are grateful to Dr Glyn Jones, the editor-in-charge of this book. It was he who contacted us and inspired us to write this book.

We would also like to thank our College chairman Arutchelvar N. Mahalingam for providing excellent facilities at Kumaraguru College of Technology, Tamil Nadu, India, where we both have been working for the past six years and teach courses on protein engineering, genomics and proteomics, molecular biology and genetic engineering. This has helped us greatly to shape the chapters in this book.

We are also thankful to our friend Sivasudha and students, especially Vidya, Arun Sathiyaseelan, Sindhu, Nikita, for critically reading our manuscript. We are also indebted to our family members and our son Navneedha Krishnan for their support at every stage of the writing of this book.

List of abbreviations

2D-PAGE	two-dimensional polyacrylamide gel electrophoresis
AFLP	amplified fragment length polymorphism
BAC	bacterial artificial chromosome
bp	base pair
cDNA	complementary deoxy ribonucleic acid
CID	collision induced dissociation
DDRT-PCR	differential display reverse transcription PCR
ddNTP	dideoxy nucleoside triphosphate
DNA	deoxy ribonucleic acid
dNTP	deoxy ribonucleoside triphosphate
DOE	Department of Energy
ELSI	Ethical, legal, social issues
ESI	electrospray ionization
EST	expressed sequence tag
FISH	fluorescent *in situ* hybridization
HGP	Human Genome Project
HUGO	Human Genome Project Organization
IEF	isoelectric focusing
IUPAC	International Union of Pure and Applied Chemistry
kbp	kilo base pair
kDa	kilo Dalton
MALDI	matrix assisted laser desorption and ionization
MALDI-ToF	matrix assisted laser desorption and ionization-time of flight
Mbp	mega base pair
mRNA	messenger ribonucleic acid
NIH	National Institute of Health
ORF	open reading frame
OTA	Office of Technology Assessment

Published by Woodhead Publishing Limited

PAC	P1 derived artificial chromosome
pI	isoelectric point
PITC	phenyl isothiocyanate
PMF	peptide mass fingerprinting
RAPD	randomly amplified polymorphic DNA
RDA	representational display analysis
RFLP	restriction fragment length polymorphism
RH	radiation hybrid
RNA	ribonucleic acid
SAGE	serial analysis of gene expression
SDS-PAGE	sodium dodecyl sulfate-polyacrylamide gel electrophoresis
SNP	single nucleotide polymorphism
SSR	simple sequence repeats
STS	sequence tagged site
ToF	time of flight
WGS	whole genome shotgun sequencing
YAC	yeast artificial chromosome

Published by Woodhead Publishing Limited

About the authors

Nachimuthu Saraswathy is Associate Professor of Biotechnology at Kumaraguru College of Technology, Tamil Nadu, India. She earned her undergraduate and Master's degree from Tamil Nadu Agricultural University, Coimbatore. She gained her PhD degree in Molecular Biology and Biotechnology from the National Centre for Biotechnology, IARI, New Delhi, in 2004. Her PhD work was devoted to studying the domain shuffling effect of BT *cry* genes and their specificity towards insects. She worked as a scientist in the Defence Research and Development Organization, at the Ministry of Defence, Government of India, studying high altitude transgenic crop development. Since 2005, Dr Saraswathy has been teaching molecular biology, genetic engineering, genomics and proteomics to undergraduate and postgraduate students of Biotechnology. She is the author of many research papers which have been published in peer-reviewed journals. She is currently carrying out a research project on bioactive wound dressing material, funded by the Department of Biotechnology, Government of India.

Ponnusamy Ramalingam is Associate Professor of Biotechnology at Kumaraguru College of Technology, Tamil Nadu, India. He gained his Master's in Biochemistry from Gulbarga University and subsequently his PhD in biochemistry from the same university. He also obtained an MTech in Biotechnology from Anna University, Chennai. Since 2001, he has been teaching an undergraduate protein engineering theory course. He is the recipient of the Young Scientist research project award from the Department of Science and Technology in New Delhi, and has received research funding to carry out research projects. He has published several research papers in national and international journals.

Published by Woodhead Publishing Limited

1

Introduction to genes and genomes

All living forms are the lineal descendants of those which lived long before.

Charles Darwin

Abstract: The **cell** is the basic structural and functional unit of all living organisms. Information governing cellular activity is stored in the nucleus of the cells. In all organisms, except in few viruses, the hereditary material is the deoxyribonucleic acid (**DNA**). In this chapter, a brief introduction to the building blocks of DNA and its structures is given. Many experiments that prove that DNA is the genetic material are discussed. Structures of prokaryotic and eukaryotic genes and genomes are also dealt with.

Key words: chromosome, DNA, gene, gene expression, genome, nucleosome, operon, ORF, replication, transcription, translation.

Key concepts

- All living organisms possess genetic material which is inherited from their parents.
- DNA is the genetic material in most organisms except in a few viruses.
- The entire genetic content present in a cell of an organism is known as the genome.
- The genome consists of coding and non-coding sequences of DNA.
- Prokaryotes have comparatively smaller genomes and are haploid. Genes of prokaryotes are organized into operons.
- Eukaryotes have large and complex genomes which are mostly diploid and sometimes polyploids. Genes of eukaryotes comprise exons and introns.
- Genetic information, which is stored in the nucleus, is used for synthesizing RNA, which, in turn, codes for a protein which executes cellular functions. This series of events is known as gene expression.

Published by Woodhead Publishing Limited

1.1 Introduction

Life on Earth originated about 3.5 billion years ago. Although many theories had been put forward to explain the origin of life on Earth, scientifically conclusive evidence came from the experiment by Miller and Urey. They found they could obtain several organic compounds resembling biological molecules from inorganic compounds by simulating the atmospheric conditions that prevailed on the primordial Earth. Primitive anaerobic self-living organisms were derived from early biomolecules, which are called coacervates. Today, there are about 1.75 million well-documented organisms in the world and it is estimated that about 30 million species are yet to be explored.

Stanley Lloyd Miller (1930–2007)

Miller, an American chemist, earned his PhD in Chemistry from the University of Chicago. He was a student of Harold Urey. He proposed a biogenesis theory and proved that biological molecules can be created from inorganic precursors through his famous Miller–Urey experiment. In fact, in 2008, it was proved that 22 amino acids could be produced using his experiment.

Harold Clayton Urey (1893–1981)

Urey, an American physical chemist, received the Nobel Prize in 1934, for his work on isotopes in deuterium (heavy water); he also played a significant role in the development of the atom bomb. Along with his student, Stanley Miller, he proposed the formation of organic molecules from inorganic compounds.

1.2 The cell

The *cell* is the basic structural and functional unit of life. In the seventeenth century, Robert Hooke discovered cells when he was observing cork, which he described as having a honeycomb-like structure. The information for life and heredity is stored in the cells of living organisms. Based on the structure of the **nucleus**, living organisms are grouped into two broad categories, the

Published by Woodhead Publishing Limited

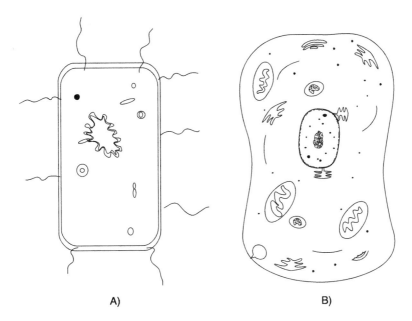

A)   B)

Figure 1.1 Prokaryotic cell (A) has simple internal structures and the genetic material is not contained inside nucleolar membrane. Eukaryotic cell (B) has complex internal structures and their genetic material is contained within defined nucleolar membrane.

prokaryotes and the eukaryotes. In prokaryotes, the genetic material is not contained in a true nucleus, as opposed to eukaryotes, in which the genetic material is stored in a separate well-organized membrane bound organelle, called the nucleus (Figure 1.1).

Robert Hooke (1635–1703)

A natural philosopher and an architect, Hooke described the law of elasticity, over which he had a dispute with Christian Huygens for a long time. Finally, he demonstrated the balance-controlled watch before the members of the Royal Society. He served in the capacity of curator of the experiment for 40 years in the Royal Society. In 1665, he published a book entitled *Micrographia*, in which he used the term 'cell' for the first time.

1.3 Mendel's contributions

All living organisms possess well-defined cellular architecture which is controlled by the **genes** that they have inherited from their parents. The branch of science dealing with heredity and variation is known as genetics. The history of genes and **genetics** dates back to Gregor Mendel's work on pea plants in the nineteenth century. Before Mendel's work, pangenesis and blending inheritance theories were the accepted theories in the biological world to explain inheritance in living organisms, but Mendel's work explained the inheritance based on scientific experiments. He is honoured with title 'Father of Genetics'. He observed that different traits such as height, colour of the flower, etc., are controlled by particulate matters present in the cell, which he termed factors. The factor controlling each trait is particulate in nature and the factors interact with each other when they are passed on to the progenies but they do not contaminate each other. In 1905, Mendel's work was rediscovered by Hugo de Vries, Carl Correns and Erich von Tschermak. Subsequently, Mendel's laws of inheritance were explained.

Gregor Johann Mendel (1822–84)

Mendel was a priest and a scientist. He devised a systematic experimental set up to study inheritance. His scientific achievements were recognized only after his death. Besides pea plants, he also worked on honeybees but he could not prove inheritance in honeybees as controlled mating was not possible.

1.4 The chromosomal theory of inheritance

Scientists were looking for the physical material of inheritance and they discovered that it is located in the **chromosomes**; this is an important step in exploring the nature of genes and its chemical make-up. Chromosomes are physical entities that can be seen and which carry the genetic material in the cell. Chromosomes are well-organized structures made of DNA and protein; since they can be stained intensely with dyes, they are called chromosomes. *Cell division* in somatic cells undergoes mitotic division, during which chromosomes duplicate and are equally distributed among daughter cells. Another type of cell division is meiosis, which occurs in germ line cells, in which the number of chromosomes in daughter cells is reduced to half.

Published by Woodhead Publishing Limited

After fertilization, the actual chromosome number of that species is restored. Careful study of movement of chromosomes during meiosis showed that chromosomes are the carriers of genetic material. Sutton and Boveri independently explained the chromosomal theory of inheritance.

Walter Sutton (1877–1916)

An American geneticist, Sutton conducted experiments on spermatogenesis on the larger grasshopper, *Brachystola magna*. He applied Mendel's law of inheritance to chromosome behaviour during meiosis and fertilization and proposed the chromosomal theory of inheritance.

Theodor Heinrich Boveri (1862–1915)

Boveri was a German biologist, whose work on sea urchins showed that all chromosomes in the cell are necessary for proper embryonic development. He also described the **centromere** involvement in cell division. He explained the existence of cell cycle checkpoints.

The **nucleosome** is the basic unit of a chromosome. A nucleosome is made up of 146 bp DNA and **histones** 2 (H2A, H2B, H3 and H4). The nucleosomes are joined in a series by the H1 linker protein. The number and size of chromosomes in a particular species never change. An arrangement of the chromosome according to their size and appearance is known as a **karyotype**. In a clinical set-up, it is usual to find chromosomal abnormalities and associated genetic disorders such as Turner's syndrome, Down's syndrome, etc.

1.5 The chemical nature of genetic material

It was established that genes are located in chromosomes. A chromosome is made up of DNA and protein. The first careful study of the nucleus was carried out by Friedrich Miescher. He isolated material from the pus cell nucleus which was rich in phosphorus and acidic in nature. He called this material nuclein, which is now known as nucleic acid. DNA (deoxyribonucleic acid) is made up of four bases, whereas proteins are made up of 20 amino acids. When compared to DNA, proteins with 20 amino acids can form more complex material in order to confer the genetic information necessary for many characters. Many people wrongly were of the view that protein was the genetic material rather than DNA, as both the protein and the DNA are present in the nucleus.

Friedrich Miescher (1844–95)

Miescher was a Swiss physician who earned his MD from the University of Basel in 1868. Due to a hearing impairment, he did not practise medicine but instead became interested in leucocytes, more specifically, in the chemistry of nucleus. He isolated material containing phosphorus, which he called nuclein.

In 1928, Frederick Griffith conducted a study on *Pneumonococcus* which exists in two forms: a non-virulent rough-type strain (II-R) and a virulent smooth-type strain (III-S). From a series of experiments, it was concluded that certain cellular material is responsible for the conversion of rough-type strains into smooth-type strains. When a rough-type strain is mixed with a heat-killed smooth-type strain, II-R is transformed into III-S. He termed this the transforming principle, but the chemical nature of the transforming principle was not yet clear.

Later, in 1944, Oswald Avery, Colin MacLeod, and Maclyn McCarty carried out Griffith's experiment in a systematic way to identify the chemical nature of the transforming principle. They conducted a series of experiments with various enzymes which are capable of destroying cellular biomolecules, such as protein, **DNA**, and **RNA** (ribonucleic acid). When the cell lysate from the heat-killed smooth-type strain of *Pneumococci* was treated with phosphatase or nucleodepolymerase, enzymes capable of degrading the DNA, the smooth-type strain was unable to transform the rough-type strain of *Pneumococci*. It was hence concluded that the transforming principle is DNA.

Even though Avery's experiment proved that DNA is the genetic material, some scientists refused to accept this principle and still supported proteins. The experiment carried out by Hershey and Chase in 1952 on *E. coli* with T_2 phage conclusively proved DNA to be the genetic material. T_2 phage is a virus which infects bacteria. It has a very simple structure and is made up of DNA and protein. Genetic material is present inside the protein coat. During infection, only DNA is injected inside the bacterial cell whereas the protein coat stays outside. Differential radio-labelling of DNA (^{32}P) and protein (^{35}S) of T_2 phage was achieved by infecting the *E. coli* strain which was grown on

Published by Woodhead Publishing Limited

^{32}P and ^{35}S radioactive media respectively. The radioactively labelled T$_2$ phage was used to infect the non-radioactive *E. coli* culture and the protein coat was separated from the infected bacteria by subjecting them to electric blending. When the non-radioactive *E. coli* was infected with ^{35}P-labelled T$_2$ phage, radioactivity was detected in the cell pellet. On the other hand, radioactivity was not observed when the ^{35}S-labelled T2 phage was used to infect the *E. coli* cells. Therefore, Hershey and Chase conclusively proved that the material injected inside the bacterial cell is DNA rather than protein and that DNA is responsible for information transfer from one generation to the next.

1.6 Composition and structure of DNA

Deoxyribonucleic acid (DNA) is made up of sugar, a nitrogenous base and a phosphate group (Figure 1.2). The combination of these molecules makes the building blocks for the DNA synthesis. The sugar present in the DNA is 2'deoxyribose, a five carbon monosaccharide, which is devoid of oxygen in its 2' position, hence the name deoxyribonucleic acid. The carbon atoms present in the *deoxyribose* are numbered 1', 2', 3', 4' and 5'. Nitrogenous bases present in the DNA can be grouped into two categories: purines (**Adenine** (A) and **Guanine** (G)), and pyrimidine (**Cytosine** (C) and **Thymine** (T)). These nitrogenous bases are attached to C1' of deoxyribose through a glycosidic bond. Deoxyribose attached to a nitrogenous base is called a nucleoside. A nucleoside attached to a phosphate group is known as a *nucleotide*. The nucleotides are linked together by phosphodiester bonds. Many nucleotides attached together are known as *polynucleotides*.

Once DNA was confirmed as the genetic material, scientists were keen to find out its chemical structure. When Wilkins and Rosalind were on the verge of solving the structure of DNA, Watson and Crick proposed their famous double helical structure of DNA in 1953 by combining the findings of Chargaff (i.e. molar concentrations of adenine equal those of guanine, and of cytosine equal those of thymine) with the X-ray diffraction data of DNA fibre. The biological form of DNA is made up of two complementary polynucleotide strands wound about each other to form a complete structure. The sugar and phosphate groups form the backbone and bases formed the core of the DNA double helix. The two strands run in opposite directions and are denoted as 5' to 3', as the ends that have the free phosphate group and the free OH groups at the terminus, respectively. Polynucleotide double-stranded DNA has a regular helix with 10 base pairs per turn. *Base pairing* is an important aspect of the DNA double helix as it helps in DNA replication and transcription.

Published by Woodhead Publishing Limited

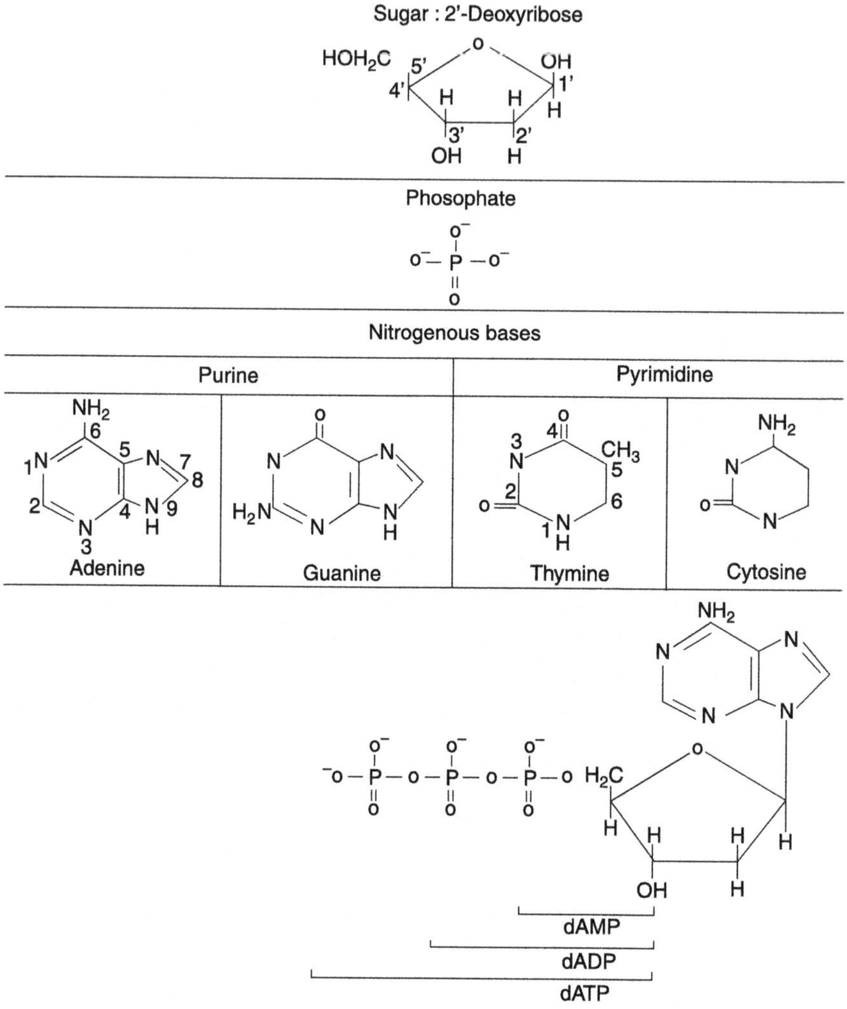

Sugar : 2'-Deoxyribose

Phosophate

Nitrogenous bases

Purine		Pyrimidine	
Adenine	Guanine	Thymine	Cytosine

dAMP

dADP

dATP

Figure 1.2 DNA is made up of five carbon sugar (deoxyribose), one phosphate group and four bases. The combination of these molecules makes the building blocks for the DNA synthesis.

1.7 The central dogma of life

The gene is the segment of the DNA strand or polynucleotide chain which determines the sequence of RNA and the subsequent protein sequence. The **central dogma of life** was established to explain the sequence of events that occur during the expression of genes.

Published by Woodhead Publishing Limited

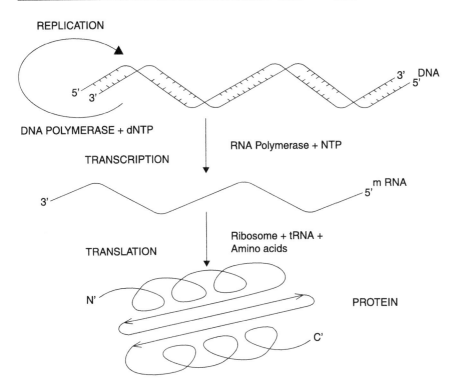

REPLICATION

DNA POLYMERASE + dNTP

TRANSCRIPTION

RNA Polymerase + NTP

TRANSLATION

Ribosome + tRNA + Amino acids

Figure 1.3 The genetic information stored in the cells replicates on its own, transcription produces mRNA and translation of mRNA produces proteins.

DNA serves as a template for RNA synthesis by the process known as **transcription** and the RNA thus synthesized serves as a template for protein synthesis by the process known as **translation**. DNA serves as template for its own synthesis by the process known as **replication**. This whole mechanism is applicable to all living organisms. This is the basic fundamental rule of life and it is called the central dogma of life (Figure 1.3).

1.8 Genomes of prokaryotes and eukaryotes

The entire hereditary information of an organism or the total DNA content of a **haploid** cell is termed the **genome**. The term genome was coined by Hans Winkler in 1920 to describe the entire genetic component of an organism. In the **prokaryotic** cells, the genome is haploid, i.e. these cells have only one set of genetic material per cell whereas **eukaryotes** have two sets of genetic material, one set is maternally inherited and the other set is paternally inherited. Hence, they are **diploid**. Some eukaryotes have more than one set of genomes; they are called **polyploid** organisms. The genome size of an organism is quantified in terms of its **C-value** and is reported in picograms (pg). Feulgen micro-densitometry, flow cytometry and DNA image cytometry are the techniques used to estimate the C-value. The C-value can be converted to the genome size in terms of base pair using the following formula:

$$\text{Genome size (bp)} = 0.978 \times 10^9 \times \text{DNA content (pg)}$$

Separate databases for genome sizes of various organisms are available on the World Wide Web. So far, C-values of 4,972 animals have been estimated; this comprises 3,231 vertebrates and 1,741 non-vertebrates. The smallest animal genome size is 0.02 pg of the *Pratylenchus coffeae* (a plant-parasitic nematode) and the largest genome size is 132.83 pg of the *Protopterus aethiopicus* (the marbled lung fish). It is observed that there is no correlation between genome sizes or C-values with the complexity of an organism (Table 1.1). This is known as the **C-value enigma**.

The genetic material of bacteria is known as **nucleoid** (nucleus-like), and within the nucleoid, the bacterial genetic material is organized along with

Table 1.1   Properties of the most common manometer liquids

Sl. No.	Scientific name	Chromosome number	Genome size (bp)	Calculated C-value (pg)*
1	ΦX174	–	5.3×10^3	0.00000542
2	*E. coli* K012	–	4.6×10^6	0.0047
3	*Haemophilus influenzae*	–	1.8×10^6	0.0018
4	*Mycobacterium leprae*	–	3.3×10^6	0.0033
5	*Bacillus subtilis*	–	4.2×10^6	0.0042
6	*Agrobacterium tumefaciens*	–	4.6×10^6	0.0047
7	*Saccharomyces cerevisiae*	2n=32	1.2×10^7	0.0012
8	*Caenorhapditis elegans*	2n=6	1.0×10^8	0.1022
9	*Drosophila melanogaster*	2n=8	1.12×10^8	0.1145
10	*Fugu rubripes*	2n=22	3.6×10^8	0.400
11	*Arabidopsis thaliana*	2n=10	1.15×10^8	0.1175
11	*Oryza sativa*	2n=24	3.9×10^8	0.3987
12	*Mus musculus*	2n=40	3.3×110^9	3.3742
13	*Homo sapiens*	2n=46	3.4×10^9	3.4764

Note: *C-value is calculated using the formula.

Published by Woodhead Publishing Limited

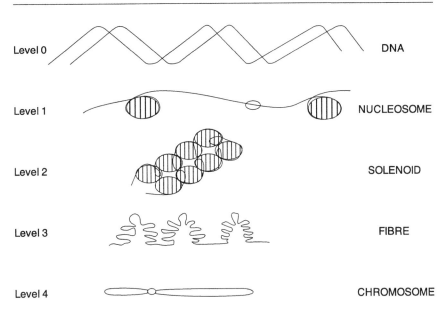

Figure 1.4 The huge DNA present in the eukaryotic cells is organized into chromosomes by following a hierarchical folding pattern.

some protein component. The bacterial genome is a single circular molecule. Some bacteria have self-replicating, extrachromosomal DNA material known as **plasmid**. Unlike prokaryotic genomes, eukaryotic genomes are well organized and larger than the bacterial genome. The **nucleus** is a membrane-bound organelle which contains the DNA and protein. The DNA is organized into several linear chromosomes by following a hierarchical folding of nucleosomes (Figure 1.4). Apart from the nuclear genome, animal cells have a mitochondrial genome which is located in mitochondria and plants have a chloroplast genome and a mitochondrial genome respectively.

1.9 The molecular structure of the gene

As we have seen already, the gene is a fragment of DNA that has the information used to code for a polypeptide. According to the central dogma of life, the information contained in the DNA fragment does not directly code for a **polypeptide**. Instead, the DNA is transcribed into RNA and the RNA acts as a template for protein synthesis. Transcription occurs in the nucleus of all organisms. RNA polymerase is the enzyme which is capable

Published by Woodhead Publishing Limited

of synthesizing RNA using DNA as the template. In prokaryotes, it is a simple process, whereas in eukaryotes, it is a complex process involving many transcription factors. Transcription of a gene occurs from a fixed position on the DNA which comprises the gene. The sequence of DNA that is represented in the RNA and codes for polypeptide is called the **open reading frame** (ORF). The sequence of DNA to which the RNA polymerase binds and which initiates transcription is termed the **promoter**. It is known as the upstream part of the gene. RNA polymerase continues to transcribe the DNA and stops at a fixed position, downstream of the gene, known as the **terminator**. These upstream and downstream regions of the gene are called the regulatory elements. The first nucleotide that is transcribed into RNA is termed the transcription start site and it is denoted as +1. Prokaryotic ORF is continuous and in most cases the structural genes involved in related metabolic activity are arranged in an **operon**. In the case of eukaryotes, the genes are little more complex. The ORF is not continuous and is split into a *coding* part or the **exon**, that is represented in the matured mRNA, and polypeptide and the *non-coding* part or the **intron**, which is not represented in the matured mRNA. Therefore, the eukaryotic genes are termed **split genes**.

1.10 Conclusion

Scientists have succeeded in identifying the genetic material that controls the cellular activity. It has been established that DNA is the genetic material in all organisms except for some viruses. The chemical composition and structure of DNA have been identified. DNA sequencing and recombinant DNA technology are used to manipulate genetic material.

Review questions and answers

1. *How did Mendel conclude that phenotypic traits are controlled by factors?*

He conducted planned crosses between pure breeding pea plants with different phenotypic traits and he observed the appearance of phenotypic traits in the subsequent generations. He then concluded that phenotypic traits are controlled by particulate factors that are now called genes.

Published by Woodhead Publishing Limited

2. *What is the composition of a nucleosome?*

A nucleosome is the basic unit of a chromosome, made up of histone core proteins 2 (H2A, H2B, H3 and H4) and 146 bp DNA wound round the core. Two nucleosomes are connected by the linker histone, H1.

3. *What is the advantage of having 2'-deoxyribose sugar instead of ribose in DNA?*

The absence of oxygen in 2' deoxyribose makes it less bulky than ribose. This helps the DNA to make a regular double helical structure.

4. *The complementarities of DNA strands help in replication and transcription. How?*

During cell division, the genetic information is perfectly passed onto the daughter cells by the replication of the DNA. During replication, the nucleotide sequence of the daughter strand is made exactly complementary to the template strand. To execute cellular functions, the genetic information stored in the DNA is passed onto RNA by the process of transcription. Thus, the complementary strand of RNA is synthesized using a DNA template.

5. *Compare prokaryotic genomes with eukaryotic genomes.*

Prokaryotic genomes are usually small (approximately 5 Mbp) and organized into single circular DNA, usually gene-rich with less repetitive DNA. Eukaryotic genomes are large (10 Mbp–100,000 Mbp) and organized into chromosomes.

Recommended reading

Avery, O.T., MacLeod, C.M. and McCarty, M. (1944) 'Studies on the chemical nature of the substance inducing transformation of *Pneumococcal* types: induction of transformation by a deoxyribonucleic acid fraction isolated from *Pneumococcus* Type III', *Journal of Experimental Medicine*, 79: 137–58.

Downie, A.W. (1972) '*Pneumococcal* transformation: a backward view, Fourth Griffith Memorial Lecture', *Journal of General Microbiology*, 73: 1–11.

Druery, C.T. and Bateson, W. (1901) 'Experiments in plant hybridization', *Journal of the Royal Horticultural Society*, 26: 1–32.

Published by Woodhead Publishing Limited

Gregory, T.R. (2005) 'The C-value enigma in plants and animals: a review of parallels and an appeal for partnership', *Annals of Botany*, 95: 133–46.

Hershey, A.D. and Chase, M. (1952) 'Independent functions of viral protein and nucleic acid in the growth of bacteriophage', *Journal of General Physiology*, 36(1): 39–56.

Miller, S.L. and Urey, H.C. (1959) 'Organic compound synthesis on the primitive Earth', *Science*, 130: 245.

Web addresses

http://data.kew.org/cvalues/introduction.html.
http://www.genomesize.com.

2

The human genome project

It is essentially immoral not to get it [the human genome sequence] done as fast as possible.

James D. Watson

Abstract: The genome denotes the entire DNA content present in a cell. Advances in DNA sequencing technology led scientists to think about the large-scale genome sequencing projects. The Human Genome Project (HGP) was one of the most ambitious and controversial biological projects ever. In this chapter, the goals, the challenges and the institutions involved in the HGP are discussed.

Key words: DOE, ELSI, genome, genome project, genomics, HGP, human genome, HUGO, NIH.

Key concepts

- The Human Genome Project (HGP) was conceived and initiated by the Department of Energy (**DOE**) in the USA in 1985. Officially, the HGP started in 1990 for a period of 15 years.
- The National Institute of Health (NIH) and the DOE took the initiative to fund the project. Along with these two agencies, 25 laboratories in five countries were also involved in this project.
- The estimated cost of the project was US$200 million per year for 15 years.
- The ultimate aim of the HGP was to determine the order of the 3 billion nucleotides that make up the human genome.
- **ELSI** related to human genome sequencing was seriously considered and 2 per cent of the money allocated was spent on this.

Published by Woodhead Publishing Limited

- Only 1–2 per cent of the human genome contains the protein coding DNA sequences; the remaining 98 per cent of the genome is made up of repeat sequences whose function is yet to be elucidated.
- The human genome consists of 23 pairs of chromosomes. Each chromosome contains euchromatic and heterochromatic regions. The DNA present in the euchromatic region is sequenced, and it is about 2.91 billion base pairs.
- Initially, the publicly funded NIH group was carrying out the sequencing of the human genome; later, a privately funded institute, Celera genomics, also started sequencing it. Both these groups reported the sequence of the human genome in 2003.

2.1 The history of the HGP

The **Human Genome Project (HGP)** is one of the largest biological projects ever in human history. It is an attempt by human beings to map their own genome. The scientific thinking on sequencing of the **human genome** has its roots in the post-Second World War effects of atomic bombs which were dropped on Japan, on Hiroshima and Nagasaki, on the 5th and 6th August 1945, respectively. Reports and scientific studies on the post-war effects on the Japanese population due to atomic radiation were undertaken. Many studies were carried out throughout the world on the survivors of the Japanese atomic bomb explosion. After many years of study it is now understood that the effects of radiation are limited not only to the current population, they have also been transmitted to the following generations. It has been conclusively shown that radiation has a major effect on the genetic make-up of living organisms. Reports show that the incidence of cancer is higher in residents of Hiroshima and Nagasaki than in the rest of the Japanese population. Therefore, many studies were undertaken on the radiation damage to the human genetic material and its ill-effects on subsequent generations.

The **Department of Energy (DOE)**, in the USA, and the International Commission for Protection Against Environmental Mutagens and Carcinogens jointly organized a meeting on 9–13 December 1984 in Alta, Utah, to discuss the serious issues related to environmental mutagens and their effect on the tiny but very important human biomolecule, i.e. DNA. The specific aim of the meeting was to develop techniques to detect mutations in the survivors of Hiroshima and Nagasaki atomic bomb explosions. Discussions were also undertaken on the possibilities of reading the human genetic material and understanding its interaction with the mutagens. This meeting ignited the idea of the HGP. A meeting was organized at the University of California, Santa

Published by Woodhead Publishing Limited

Cruz, under the chairmanship of well-known molecular biologist, Robert L. Sinsheimer in 1985, which also discussed the possibility of sequencing the human genome. In 1986, Charles DeLisi, a biophysicist and administrator of the Health and Environmental Research Committee (HERAC) of the DOE, convened a meeting on the possibility of sequencing the human genome and transformed the idea into reality by proposing a three-point programme which included genome mapping, sequencing, and computer-aided arrangement of the sequences. He prepared a document entitled 'Human Genome Initiative' and submitted it to the Office of Technology Assessment (OTA) for approval. After a year-long debate on various aspects, the National Research Council of the USA recommended a multi-phased research project. In 1988, the US Congress approved the HGP and the fund was allocated through the NIH and DOE budget. The HGP officially started in 1990 and the duration of the project was fixed for 15 years. The International Human Genome Sequencing Consortium (IHGSC) was established to coordinate the HGP operations in different institutions throughout the world.

Robert Luis Sinsheimer

American biophysicist, born in Washington, DC. He obtained his undergraduate degree in Quantitative Biology and a PhD in Biophysics from Massachusetts Institute of Technology. He served as the Chancellor of University of California, Santa Cruz, from 1977 to 1987. He chaired the meeting on the benefits of Human Genome Sequencing and this meeting formed the basis for the HGP. He received a Presidential Citizen Medal for his contribution to science in 2001.

Charles DeLisi

A cancer biologist, born in New York, DeLisi obtained his PhD from New York University. After working in many college campuses in the USA, he joined DOE as the Director of Health and Environmental Education from 1985 to 1987. He received a Presidential Citizen Medal for his contributions to the HGP.

Map-based cloning sequencing strategy was followed by the publicly funded project. This method is known as **BAC**-based method or **clone-by-clone sequencing** method. In 1998, a privately funded organization led by Dr Craig

J. Venter also started human genome sequencing using a separate strategy in which sequencing was done without preparing a high density map, and this is known as the 'shotgun method' of sequencing. It became a competition between the publicly funded HGP and the privately funded HGP to complete the human genome sequence. But later both agencies agreed to work together and jointly announced the completion of the human genome sequence in 2003. The HGP was completed two years ahead of the planned schedule.

The HGP was initially headed by J. D. Watson and later headed by Francis S. Collins, the Director of the NIH. The draft sequence was released in 2000 and the completed sequence was released in 2003. Subsequently, the IHGSC finished genome sequencing with many gaps and the completely assembled human genome sequence was released, covering 99 per cent of the euchromatic region. With an error rate of 1 error per 100,000 bases, 2,851,330,913 nucleotides were sequenced. The human genome sequence is deposited in NCBI Human Build 35 denoted as NC000001–NC000024.

2.2 The budget for the HGP

When the human genome project was aired, many scientists argued that it was unnecessary to sequence the entire genome as it involved a lot of money. Therefore, specific objectives were developed which involved the most efficient techniques which would cost less. The initial estimate for the HGP was US$200 million per year for 15 years. Rapid progress was made in mapping and sequencing techniques which paved the way for the completion of the HGP ahead of its target schedule 2005. Therefore, the project cost was less than what had been expected. The budget of the HGP was only approximately US$2.7 billion instead of US$3.0 billion. About 50 per cent of the fund was provided by the USA and the rest came from other countries, mainly the UK, France, Germany, Japan, Australia, and Canada. In 1988, the Office of Technology Assessment sanctioned funds for the HGP through the NIH and DOE, which were US$17.3 million and US$11.8 million, respectively.

2.3 Goals of the HGP

The idea behind the HGP was to learn the genetic mechanisms that control the human development from zygote, its interactions with the environment, genetic disorders, the aging process, etc. It is impossible to understand all these in a single experiment. Therefore, specified goals for the HGP were established which covered not only the issues which were related directly to

Published by Woodhead Publishing Limited

the human being but also addressed other related problems. When the HGP started, available mapping and sequencing technologies were not advanced enough to sequence the 3 billion bases of the human genome in the specified time frame. They were costly, slow, and not accurate. Therefore, specific goals were set to increase the efficiency. The goals of the HGP were divided into three five-year plans before proceeding to the next five-year plan; the progress and hurdles were analyzed and suitable modifications were made. The specific objectives of the HGP were jointly prepared by the National Institute of Health and the Department of Energy for three five-year periods, 1991–95 (FY Plan No. 1), 1996–2000 (FY Plan No. 2) and 2001–03 (FY Plan No. 3).

Goals of the Human Genome Project

1. To prepare a high resolution map of the human genome using genetic as well as physical mapping techniques.
2. To determine the order of the nucleotide arrangement in all the 22 autosomes and the two sex chromosomes (X and Y).
3. To develop high throughput sequencing technology.
4. To learn the sequence of genomes of model organisms to test the feasibility of different mapping and sequencing techniques.
5. To develop computer tools to store the sequence data and access the stored sequence information for various purposes.
6. To annotate the DNA sequence based on its sequence content, such as ORF, promoter, terminator, enhancers and repeat sequences present in the genome.
7. To address ethical social and legal issues that might arise pertaining to human genome sequencing and its use.

2.4 Laboratories and investigators involved in the HGP

The Human Genome Project is a multinational project that involved investigators from international research laboratories. Many prominent scientists from 20 laboratories located in different countries were involved in developing techniques and sequencing. Specific centres were identified to conduct different aspects of the HGP. In the USA, the DOE and the NIH coordinated the sequencing programme. Under the guidance of these two agencies, scientists working in national laboratories, universities, clinical

Published by Woodhead Publishing Limited

centres, and private research centres worked on the HGP. Subsequently, other countries also started sponsoring the human genome sequencing project and contributed to the efforts. The UK contributed about one-third of the fund of the HGP. The Medical Research Council of the UK, a government-sponsored institution, supported the HGP. Other non-governmental agencies like the Wellcome Trust and the Imperial Cancer Research Council also took an active part in the HGP. In France, the Centre d'Etude du Polymorphisme Humain (CEPH) in Paris, which has the largest collection of human family cell lines from all over the world, started working on human genome mapping under the guidance of Daniel Cohen and Jean Dausset.

2.5 The Human Genome Organization (HUGO)

The HGP is an international effort, and scientists from more than 18 countries are involved in this project. To ensure the efficient functioning of this multinational, multi-phased project, coordination was important. A separate agency called the Human Genome Organization (**HUGO**) was established in 1989. The main objective of agency was to coordinate the sequencing project in order to avoid unnecessary competition among scientists, to avoid duplication of work and also to encourage the exchange of scientific material and data relevant to genome sequencing among scientists. It acted as the coordinating agency to conduct training programmes, and as nodal agency to provide information relevant to genome sequencing to the public, and to solicit opinions so as to address the ethical, legal, and social issues (ELSI) related to genome sequencing. It is also involved in providing fellowships, training, and course materials related to genome sequencing. It provided expert advice to governments on developments in genome sequencing. Meetings and workshops were conducted in a phased manner in different locations to exchange ideas and materials related to HGP.

2.6 Salient findings of the HGP

1. It was found that only ~2 per cent of the human genome is made up of coding DNA. The number of genes predicted (20,000–25,000) is far fewer than anticipated (100,000) and comparatively less than the earlier prediction from a rough draft (31,000). It was found that on average one gene codes for 2–3 proteins through the alternate splicing method.

Published by Woodhead Publishing Limited

2. The proteome of a human is ten times larger than that of a fruit fly or a worm. It has also been found that the finished genome showed fewer **exons** per transcript (4.7 exons per transcript) and shorter ORF (847 amino acids) than earlier predictions, i.e. 9.7 exons per transcript and 1,487 amino acids.

3. The number of genes present in a human is two to three times more than that of the tiny fruit fly (13,500) and the worm (19,000).

4. Comparison of the entire human genome sequence with other model organism sequences showed that many of the human genes share a sequence similarity with genes present in other organisms. These genes are called **orthologous genes,** and are medically important. This will be useful in understanding the functions of genes in other model organisms.

5. Regulation of gene expression is an important aspect in eukaryotes. Many of the genes code for transcription factors which interact with **enhancers** and silencers to give flexible gene expression.

6. Genes are not distributed equally throughout the chromosome; a particular chromosome contains gene-rich and gene-poor regions. All chromosomes do not have a similar number of genes, for example, the human chromosome 19 has the largest number of genes while chromosome 5 has the least.

7. **Segmented duplication** is a common phenomenon observed in all chromosomes. Segmental duplications are regions on chromosomes with more than 1 kbp length, having more than 90 per cent sequence identity. It has been observed that segmental duplication is more common in humans than in mice, and chromosome Y has a maximum of 25 per cent of its length composed of segmental duplication.

8. Repetitive DNA accounts for major part of the genome which varies from short repeats to long repeats. About 85 per cent of the heterochromatic region is not sequenced yet, because these DNA fragments are difficult to maintain as clones and also, after sequencing, it is difficult to arrange them in order.

9. The total length of coding sequences in the human genome sequence is approximately 34 Mbp and it accounts for about 1.2 per cent of the euchromatic region, and 0.7 per cent of euchromatic sequences are untranslated.

10. From the current human gene catalogue (Ensembl 34d), 22,287 gene loci comprising 19,438 known genes have been identified and 2188 have been predicted.

Although the ultimate goal of the HGP was to sequence the entire genome of humans, this was not achieved. This is due to the fact that the chromosomes are made up of two kinds of sequences: the **euchromatic** region which is rich in genes and the **heterochromatin** region present in the **telomere** and **centromere** regions, which is rich in repeat sequences and is devoid of genes.

2.7 Potential applications of the HGP

Life Science has become a rewarding research area in the twenty-first century. It is estimated that biotechnology-related products will account for US$50 billion in 2011. The HGP not only provided the genome sequence, it also provided a wealth of information and technologies that can be utilized in various areas. Complete sequencing of human genome and the genomes of model organisms gave a complete picture on the number of genes present in them. The potential benefits of the HGP are now described.

Pharmaceuticals related to DNA-based products are projected to grow every year. The human genome source book is going to be an important reference for biomedical research. Analysis of the finished human genome sequence showed that 5.3 per cent of the euchromatic regions carry segmental duplications; these segmental duplications are of medical interest because they are responsible for deletion and rearrangements of chromosomal regions which result in many genetic diseases like Williams syndrome, Charcot-Marie-Tooth region, DiGeorge syndrome region, etc. It can also be used to develop highly sensitive techniques to detect genetic diseases. High resolution genetic maps as well as physical maps were also developed for many organisms, including humans. These can be used to identify the predisposition of humans to many diseases like diabetes, cancer, etc. It has been estimated that approximately 4,000 diseases are associated with mutations in various genes. Once the genes involved in a metabolic pathway have been correctly elucidated, the missing gene product can be identified. If the structure of a particular protein is available, drugs can be developed to correct the non-functional proteins.

The response of an individual to a drug varies with their genetic make-up. If patients with a similar disease are given the same dosage of a drug, not all of them show the same response. Negative responses are also observed sometimes. Once the whole genome sequence is available, the drug dosage for each individual patient can be optimized. This is known as personalized medicine. Graft rejection is a major problem faced during organ transplantation. Graft rejection involves many genes. The HGP provides necessary information to identify all the genes involved in this process, and, thus, organ matching can be done perfectly before proceeding to organ transplantation.

2.8 Post-HGP challenges

After sequencing the human genome, this created a great hype among scientists across the world. Many scientists have suggestions on how to utilize this enormous information for a variety of purposes. To make this information useful for biomedical research, further analysis of the sequence is needed. The data generated from the human genome sequence and other genome sequences are stored in the form of nucleotide data. The nucleotide sequence alone yields no information. In other words, sequencing a large segment of DNA will yield a sequence of letters with no obvious information. The hidden information has to be understood and the possible functions have to be elucidated. This process is known as **genome annotation**. This is the most important post-HGP challenge.

2.9 Ethical, legal, social issues (ELSI) related to the HGP

Since the inception of the HGP, concerns have been expressed about the possible use of the human genome sequence information. About 3–5 per cent of the HGP fund was allocated to study the ELSI related to the HGP. Policy issues were discussed related to disclosure of personal genome sequence information to the public, and the storage and use of genome information for different purposes. In May 2008, the US President Bush signed a Bill called the Genetic Information Nondiscrimination Act (GINA), which states that insurance companies and employers should not use genetic tests to discriminate between people. Because of this law, the people of the United States of America can undergo genetic tests freely to discover if they have genetic diseases. One of the benefits of the human genome sequencing project is that the information can be used to identify the genes causing genetic diseases in humans. There is no well proven treatment for many such diseases. Being tested positive for genetic diseases for which no treatment is available creates anxiety and trauma for the patients. Another important ethical view on the HGP is the end applications of the sequence obtained. With the spectacular advancements made in genetic manipulation techniques, it is now possible to manipulate the genome of a person to rectify the mutated gene through somatic gene therapy, which is ethically acceptable. Also, genetic interventions can be made to enhance a particular trait which may be socially good or bad.

Published by Woodhead Publishing Limited

Timeline of the HGP

1984: DOE convened a meeting in Utah to discuss and develop sensitive techniques to detect mutations in survivors of the Japanese atomic bomb explosion in 1945.

1985: A meeting was organized at the University of California, Santa Cruz, under the chairmanship of the well-known molecular biologist, Robert L. Sinsheimer, to discuss the possibility of sequencing the human genome. Simultaneously another meeting was held in Santa Fe by Charles Delisi and David Smith on the Human Genome Initiative.

1986: Charles DeLisi, a biophysicist and administrator of the Health and Environmental Research Committee (HERAC) of the DOE, convened a meeting on the possibility of sequencing the human genome and transformed the idea into reality.

1988: The National Research Council in the USA recommends the multi-phased research project. HUGO is formed. The NIH and the DOE sign an MoU on the HGP.

1989: The NIH and the DOE jointly set up ELSI.

1990: The HGP started to map the human genome and prepare **BAC libraries**.

1991: Setting up of human chromosome data repository.

1992: Human genome genetic map was completed.

1993: International IMAGE consortium formed to coordinate efficient mapping and cDNA sequencing techniques.

1995: Completion of the smallest bacterial genome *Mycoplasma genetlaium*, the whole genome of *H. influenzae* completed. High resolution physical map of human chromosomes 16 and 19.

1996: First Archaeal genome of *Methanococcus jannaschii* completed, *Saccharomyces cerevisiae* completed. Sequencing data release policy was established at the Bermuda conference.

1997: High resolution physical map of human chromosomes X and 7.

1998: *Caenorhabdidtis elegans* and *Mycobacterium tuberculosis* genomes sequenced. Celera Genomics started to sequence the human genome using the shotgun sequencing methods.

1999: First human chromosome 22 completely sequenced. One billion base pairs of human genome were completed.

Published by Woodhead Publishing Limited

2000: Draft sequence of human genome was published. Smallest human chromosome 21 was completed, *Drosophila melanogaster* completely sequenced.

2001: Chromosome 20 was sequenced.

2002: Draft sequence of mouse and *Fugu rubripes* genome was completed.

2003: Human genome sequencing completed, Celera and public funded project announced human genome sequence together.

2004: The number of genes present in human genome found to be 20,000–25,000.

2005: Gaps were closed in human chromosomes 2, 4 and X.

2006: Gaps were closed in human chromosomes 1, 3, 11, 12, 15, 17 and 18.

2007: First diploid human genome sequence was published.

2008: Genetic Information Nondiscrimination Act (GINA) becomes law in the USA.

2009: Third generation DNA sequencing technologies published.

2.10 The international HapMap Project

This is an international project to identify the genes responsible for human diseases and drug-response genes present in the human being. This project was started in October 2002; six countries (the USA, the UK, Japan, China, Canada and Nigeria) were involved in this project. One of the main objectives of the HapMap Project is to find **haplotypes** in human population. Haplotypes are regions of DNA that are shared by multiple individuals of a population. Numerous studies indicated that common diseases like cancer, heart disease, diabetes, depression and asthma have common genetic variants. Therefore, the HapMap project is expected to give information on links between the occurrence of these diseases and the genes responsible for them. Blood samples were collected from 270 people from these countries and converted to cell lines by the Coriell Institute of Medical Research, New Jersey, USA, a non-profit organization supported by the NIH which supplies the cell lines to the researchers of the HapMap project. Some 3.1 million SNPs have been accurately mapped in the human population and released in dbSNP, which is freely available for researchers for further work on them.

Review questions and answers

1. *What is the role of the Department of Energy in the HGP?*

The Department of Energy (DOE) has been involved in DNA research for a long time. Their main idea was to develop techniques to detect mutations and their consequences in the human genome. Eventually, when the meeting was conducted in 1985 in Utah with the International Commission for Protection against Environmental Mutagens and Carcinogens, a conclusion was reached. The DOE could put forward the proposal to sequence the human genome.

2. *List the goals of HGP.*

- To prepare a high density map of the human genome.
- To determine the sequence of 3 billion nucleotides of the human genome.
- To develop high throughput sequencing techniques.
- To store and analyze the large collection of DNA sequences and maintain them in a database which is accessible to all.
- To address the ethical, social and legal issues (ELSI) related to the Human Genome Project.
- To train scientists all over the world in genomics technologies.
- To annotate the genome sequence in terms of ORF, regulatory elements, enhancers, repeat sequences, etc.

3. *What are the post-genome sequencing challenges?*

Genome sequencing does not end with just determining the order of nucleotide present in a genome. To gain useful information from it, all the genes and their functions have to be identified.

4. *List the spin-off technologies developed through HGP.*

- High-throughput sequencing technologies have been developed. Fully automatic robotic DNA sequencers with a capacity of 1 million nucleotides per day are possible at low cost and with high accuracy.
- Genome sequences of many model organisms were completed. Comparative genome analysis with these organisms and the human genome will give the functions of orthologous genes across a genus.
- Bioinformatics tools and databases were developed.

5. What is the relevance of ELSI in the context of HGP?

About 2–5 per cent of the HGP fund was allocated to study the ELSI related to the HGP. ELSI addresses the ethical issues related to genetic information access, such as maintaining confidentiality, the risk associated with genetic testing of a personal genome, etc.

6. List the salient findings of the HGP.

- High density genetic and physical map of human genome has been prepared.
- Only 2 per cent of the human genome represents the protein coding genes. Thus 98 per cent of the human genome is made up of repeat sequences.
- Highly repetitive sequences located in heterochomatic regions are not sequenced as they are not amenable to sequencing.
- The number of genes present in the human genome is less than expected. It is predicted to be 22,287 as against 35,000 genes which were predicted from the draft human genome sequence.

Recommended reading

Bentley, D.R. (1996) 'Genomic science information should be released immediately and freely in the public domain', *Science*, 274: 533–4.

Collins, F.S., Morgan, M. and Patrinos, A. (2003) 'The Human Genome Project: lessons from large-scale biology', *Science*, 300: 286–90.

Dausset, J., Cann, H., Cohen, D., Lathrop, M., Lalouel, J.M. and White, R. (1990) 'Centre d'étude du polymorphisme humain (CEPH): collaborative genetic mapping of the human genome', *Genomics*, 6: 575–7.

International Human Genome Sequencing Consortium (2001) 'Initial sequencing and analysis of the human genome', *Nature*, 409: 860–921.

International Human Genome Sequencing Consortium (2004) 'Finishing the euchromatic sequence of the human genome', *Nature*, 431: 931–45.

Pankhurst, R.C. and Holder, D.W. (1952) *Wind tunnel technique*, London: Pitman.

Robert, L., Davenport, R.J., Pennisi, E. and Marshall, E. (2001) 'A history of the Human Genome Project', *Science*, 291(5507): 1195.

Schull, W.J. (1995) *Effects of Atomic Radiation: A Half-Century of Studies from Hiroshima and Nagasaki*, New York: Wiley-Liss, Inc.

Speaker, M.S.L. and Hansen, E. (2003) *A Guide to the Human Genome Project Technologies, People, and Institutions*, Pennsylvania: Chemical Heritage Foundation.

Venter, J.C., et al. (2001) 'The sequencing of the human genome', *Science*, 291: 1304–51.

Web addresses

http://genome.ucsc.edu/.
http://snp.cshl.org/thehapmap.html.en/.
http://www.ncbi.nlm.nih.gov/projects/genome/guide/human/.
http://www.ncbi.nlm.nih.gov/projects/SNP/.
http://www.ornl.gov/sci/techresources/Human_Genome/elsi/elsi.shtml.
http://www.ornl.gov/sci/techresources/Human_Genome/home.shtml.

3

Genomes of model organisms

Abstract: Model organisms play a major role in understanding human physiology from different angles. In this chapter, the importance and essential features of genetic model organisms are presented. Most commonly studied model organisms for genomic studies are discussed under different groups such as virus, bacteria, fungi, insect, plant and animal. In each category one particular organism is discussed in detail.

Key words: *Arabidopsis* genome, bacterial genome, *Drosophila* genome, model organism, mouse genome, rat genome, rice genome, viral genome, worm genome.

Key concepts

- Genetic make-up and basic life processes like replication, transcription and translation are the same in all organisms.
- Experimental results from one organism can be extrapolated to another organism.
- Genome size correlates approximately to the complexity of the organism. Viruses and bacteria have small genomes, plants and animals have larger genomes.
- The complexity of the organism increases as the size of the genome increases.
- Biological studies were successfully conducted on certain organisms and these organisms are called model organisms.
- For each group of organisms, one particular species is selected and used as a model organism.

3.1 Introduction

Life on Earth is based on the simple concept of unity in diversity. Biological processes such as genetic make-up, biochemistry, physiology and reproduction in living organisms share a great degree of unity at the same time as they maintain a certain level of diversity in terms of immune system, phenotypic appearance, etc. Advances in biology are due to data obtained from studies on **model organisms**. A model organism should be simple, easily accessible and should also have some special features. There are many unique features that make an organism suitable to be considered as a model organism for biological studies. A model organism is selected with the specific objective of its usefulness in understanding human biology and any subsequent applications in human healthcare. Studies of model organisms are helpful in understanding the underlying common evolutionary relationship between them. There are different types of model organisms such as genetic models, experimental models and genomic models. Genomic model organisms were selected based on the nature and size of the genome and previous genetic studies in that organism.

Reasons for working with model organisms

1. A particular biological phenomenon is easier to understand when studied in a model organism than when it is studied in humans.
2. Easy genetic manipulation.
3. Availability of genetic information such as the genetic map, the physical map and markers for mapping.
4. Short life cycle.
5. High fecundity.
6. Potential applications in human healthcare.
7. Lower cost of study.
8. Ease of maintenance.
9. Less of an ethical problem.
10. Share a certain degree of relationship with human physiology.

3.2 The viral genome

The existence of a virus was first detected in plants by a Russian botanist, Ivanosky, in 1852. The tobacco mosaic virus (TMV) was the first virus to be described for its infectious nature on tobacco leaves. Later, it was found that viruses are ubiquitous intracellular parasites. The importance of viruses in human life is due to their pathogenicity on humans, animals, plants and beneficial bacteria. Even 150 years after their discovery, there is no proper control measure for many of the human viral diseases. Viruses are smaller than bacteria and therefore they are difficult to observe under an ordinary light microscope. The detailed structure of viruses was elucidated only after the electron microscope had been invented. The unique characters of viruses include lack of cellular structures and the ability to live outside the cell without any life-associated activities.

A virus starts its life only when it encounters a living cell; until then it is present in the environment as an inert material. The process of attachment of a viral particle to the host cell is known as landing or adsorption. The viruses enter the host cell through many different modes, the most common method being interaction of viral receptor and cell surface receptors. After successful attachment, either the entire viral particle is ingested into the host cell by **endocytosis** (an animal virus), or it sends only nucleic acid inside the cell (a **bacteriophage**). The viral proteins are successful in controlling the host cell machinery. Viruses have either DNA or RNA as their genome, according to the structure of the genome; the genomes of viruses are classified as dsDNA genome, ssDNA genome, dsRNA genome and ssRNA genome. Depending upon the nature and structure of a viral genome, it adopts various modes of replication. The genomes of viruses are circular, linear or segmented. The replication of the viral genome can be divided into two stages: (1) early stage: proteins needed for viral genome are synthesized and many copies of the viral genome are produced; and (2) late stage: proteins needed for viral assembly are synthesized. Fully matured viral particles are released from the host cell as progenies. Viral genomes are small and are easy to handle and manipulate. The first viral genome sequenced was ΦX173.

Lambda phage

Lambda	Group I (dsDNA)
Order	Caudovirales
Family	Siphoviridae
Genus	Lambda-like virus
Species	**Lambda phage**

3.3 Bacterial genomes

Bacteria are single cellular organisms, found in every environment on Earth. There are two classes of bacteria, namely, eubacteria and **archeaebacteria**. A bacterial genome is usually circular although a linear bacterial genome has also been reported. In bacteria, the genome is not contained inside a nucleus, rather it is present as naked DNA molecule and is called a **nucleoid**. Therefore, they are called **prokaryotes**, i.e. cells with primitive nucleus. Evidence for multiple and linear chromosome present in bacteria came from **pulse field gel electrophoresis (PFGE)**. The structure of the bacterial genome has been demonstrated using an electron microscope for both Gram negative (*E. coli*) and Gram positive bacteria (*Bacillus subtilis*). Many bacterial cells are characterized by the presence of extra-chromosomal DNA called **plasmid**. In total, 1241 bacterial genomes have been sequenced and deposited in EMBL-EBI. **BacMap** is an interactive visual database containing hundreds of fully labelled, zoomable, and searchable maps of bacterial genomes.

3.3.1 *Escherichia coli* genome

One of the best-known model organisms for prokaryotes, specifically for bacteria, is *E. coli*. It lives in the lower intestine of animals, including human beings. Although *E. coli* are considered non-pathogenic, certain serotypes are food-borne pathogens. This was first described by Theoder Eschesrich in 1885. *E. coli* is a Gram negative, non-spore-forming, facultative anaerobe. Optimal growth condition occurs at 37°C. Non-pathogenic *E. coli* strains are used as probiotic. *E. coli* K12 is a laboratory strain and is used for molecular biological studies; it has become a popular model organism due to its rapid growth rate and simple nutritional requirements. It is capable of transferring genes horizontally through **transduction** and **conjugation**. The **genome map** of *E. coli* was mapped for the first time using the conjugation technique. The complete **genome sequence** of *E. coli* was reported in 1997 by Blattner et al., at the University of Wisconsin. The salient features of the *E. coli* genome are single circular chromosome of size 4,639,221 bp, the G+C content of the genome is 50.8 per cent and nearly 88 per cent of the genome codes for 4288 proteins, of which many are annotated with specific functions. About 0.8 per cent of the genome represents genes-coding for rRNA, tRNA, etc., and 0.7 per cent of the genome is made up of repeat sequences and the remaining 11 per cent of the genome harbours other regulatory sequences. Since it shares a high degree of genotypic and phenotypic similarities with other pathogenic bacteria and also exists inside the human body, the host–pathogen interaction can be well studied using information from the *E. coli* genome.

Published by Woodhead Publishing Limited

Escherichia coli (E. coli)

Kingdom	Bacteria
Phylum	Proteobacteria
Class	Gamma proteobacteria
Order	Enterobacteriales
Family	Enterobacteriaceae
Genus	Escherichia
Species	*Escherichia coli (E. coli)*

Salient features of *E. coli*

1. The *E. coli* genome was sequenced in the University of Wisconsin in 1997 by F. Blattner et al.
2. The total genome size is 4,639,221 bp and the total number of genes predicted is 4403.
3. The universally accepted laboratory model bacterium is *E. coli* strain K-12.
4. Both harmful and beneficial strains of *E. coli* help us understand their effect of humans.
5. It is simple to grow and maintain.
6. Horizontal gene transfer between bacteria was first demonstrated in *E. coli*.
7. It is used as an indicator organism for environmental foecal contamination.
8. It serves as natural host for many bacteriophages which are used for gene expression studies and cloning vectors.
9. Many cloning and expression vectors for gene manipulation have been developed.
10. Fundamentals of DNA replication, transcription and translation were studied using this organism.

3.4 Fungal genomes

3.4.1 Yeast: *Saccharomyces cerevisiae*

Saccharomyces cerevisiae has been used in the brewing industries since ancient times. Originally, it was isolated from grape skin. It is also called

Published by Woodhead Publishing Limited

budding yeast or baker's yeast. Many human genes which are homologous to yeast genes have been identified with respect to cell division, cell signalling, protein processing, etc. Yeast cells exist in two forms: **haploid** and **diploid**. Haploid cells grow and undergo mitosis division. Under stressed conditions, diploid cells undergo meiosis and are converted into haploid cells for further growth. The haploid cells exist in two mating types and show primitive sex differentiation. This is used as the model organism to study the cell division. Yeast grows aerobically on a medium containing sugars like glucose, trehalose and maltose and ammonia and urea as nitrogen sources.

The genome of *S. cerevisiae* was the first eukaryotic to be genome sequenced. The complete sequence of the yeast is available from 1996 onwards. The total genome size is 12,156,677 bp and the number of genes identified is 6275, organized into 16 chromosomes. It shares about 23 per cent with human genes. The *Saccharomyces* Genome Database (SGD) is an important site for all information regarding *S. cerevisiae*.

Saccahromyces cerevisiae

Kingdom	Fungi
Phylum	Ascomycota
Class	Saccharomycetes
Order	Saccaharomycetales
Family	Saccharomycetaceae
Genus	Saccharomyces
Species	*S. cerevisiae*

Salient features of yeast

1. Cell division is similar to that of humans and is regulated by homologous gene products.
2. Sex differentiation exists under haploid condition.
3. DNA repair and DNA damage pathways are well studied.
4. Single cell organism with short doubling time of about 1.5 hours to 2 hours.
5. Gene knockout through homologous recombination is well established.
6. It has a high economic potential in the fermentation industry.
7. First eukaryotic genome sequenced.

3.4.2 *Neurospora crassa*

This is a filamentus fungi commonly called bread mould, found to grow in the natural environment and on dead plant matter. It is a multicellular filamentous fungi having medicinal, industrial and agricultural importance. *Neurospora* was known to scientists when it was used to prove the one-gene, one-enzyme hypothesis by Beadle and Tantum in 1946. Being a multicellular fungus, it was explored for cellular differentiation and the sexual cycle. It is a good model to study epigenetic mechanisms. Repeat Induced Point (RIP) mutation is one such phenomenon by which a particular genome protects itself from mobile genetic elements. *Neurospora* is a saprotroph; its genome contains a lot of genes which match the plant's pathogenic fungal genes. Therefore, the genome sequence of *Neurospora* can be used to identify pathogenesis-related genes in other fungi.

The complete genome sequence of *Neurospora* was reported in 2003. The characteristic feature of *Neurospora* genome is the size of the genome, which is a 43 million base pair, and that it codes for 10,000 genes in seven linkage groups. *Neurospora* genome was sequenced using whole genome shotgun sequencing, using the paired end sequencing method. It has seven chromosomes with G+C content of about 50 per cent with very less repetitive DNA. The average gene length is 1.67 kb and the gene density is one gene per 3.7 kb.

Many discoveries were made using the *Neurospora* genome sequence such as red light photobiology, Ca^{2+} signalling, etc. It shares more animal-specific characters than yeast, such as circadian rhythms, gene expression control by DNA methylation, post-transcriptional gene silencing and a primitive respiratory chain, etc. The number of genes predicted from *N. crassa* genome is 10,082 which is double the number of genes than predicted from *S. pombe* and *S. cerevisiae*.

Neurospora crassa

Kingdom	Fungi
Phylum	Ascomycota
Class	Ascomycetes
Order	Sodariales
Family	Sardariceae
Genus	Neurospora
Species	*N. crassa*

Published by Woodhead Publishing Limited

Salient features of *Neurospora crassa*

1. The famous concept of molecular biology, i.e., one gene, one enzyme, was proved in this organism for the first time.
2. It is the simplest organism to display Mendelian genetics and can be easily grown on microbiological media.
3. It has a genetically tractable life cycle.
4. It follows an asexual or sexual cycle depending on the surrounding medium.
5. It was the first multicellular filamentous fungus whose genome was sequenced.

3.5 Worm genome: *Caenorhabditis elegans*

C. elegans was introduced by Sydney Brenner in 1963 as a model organism specifically to study developmental biology and neurobiology. It is a multicellular eukaryote and shares a cellular organization such as gut, excretory pathway, neural network, etc., similar to other higher eukaryotes. It has five systems: digestive system, nervous system, sensory system, muscular system and reproductive system. It has a short life cycle; egg to egg is only 2 to 3 weeks. It feeds on *E. coli* and therefore can be easily maintained in a Petri dish on an *E. coli* culture. The developmental pathways of all 957 cells are marked. It has six pairs of chromosomes (2n = 6) and the ratio of autosomes to sex chromosomes determines the sex of *C. elegans* as male or hermaphrodite. Because of self-fertilization of hermaphrodites, 300 progenies are produced with a uniform genetic identity. It has a comparatively smaller genome (97 Mbp) than humans but 35 per cent of the worm genes have a human homolog. It is well known in gene function studies using the RNAi technique. It is the only organism in which all the synaptic junctions have been mapped. *C. elegans* also has medically important genes. It shares many ailments like cancer, neuro-degeneration, spinal muscular atrophy, juvenile Parkinson's disease and many others with humans. Owing to its transparent body, the mutant phenotype can easily be identified. It is one of the smallest and simplest animals with a nervous system.

The genome of *C. elegans* is approximately 100 million base pairs which codes for about 20,000 genes, of which 6000 genes are human homologs. Its genome is 20 times larger than the *E. coli* genome and it is only 1/30 of the human genome. It is the first animal genome sequenced; the complete sequence has been available since 1998. Clone-by-clone strategy was

Published by Woodhead Publishing Limited

followed for full-length genome sequencing. Two important centres involved in *C. elegans* genome sequencing are the Washington University Genome Sequencing Center (WUGSC) supported by the National Human Genome Research Institute (NHGRI), USA, and the Sanger Institute, supported by the Wellcome Trust, UK. Information regarding map, sequence and phenotype classification can be accessed through a repository specifically devoted to *C. elegans* called **WormBase**. WormBase is a community database which provides all information regarding *C. elegans* and related species. The entire worm genome is divided into 3,000 clones and each can be accessed through WormBase.

In 2002, the Nobel Prize was awarded to scientists working on *C. elegans*: Sydney Brenner for initiating works on *C. elegans*, John Sulston for cell lineage mapping and Robert Horvitz for studying programmed cell death. In 2006, another Noble Prize was awarded to Fire and Melo for RNAi in *C. elegans*.

Sydney Brenner

Brenner was born on 13 January 1927 in South Africa, and was awarded the Nobel Prize in physiology or medicine in 2002 for his research on *C. elegans*. He is one of the lucky scientists who saw the DNA helix model prepared by Watson and Crick. Later he worked with Francis Crick and proposed the concept of transfer RNA.

Caenorhabditis elegans

Kingdom	Animalia
Phylum	Nematoda
Class	Secernentea
Order	Rhabditidae
Family	Rhabditidae
Genus	*Caenorhabditis*
Species	*C. elegans*

Salient features of *C. elegans*

1. It is a non-parasitic soil nematode and its name liteally means recent rod-like nice animal.

Published by Woodhead Publishing Limited

2. Use of this animal as a model organism is due to its transparent body which helps in easy observation.
3. Size is small (approximately 1 mm length and 80 μm width).
4. A large number can be grown in a Petri dish (approximately 10,000 worms per Petri dish).
5. Since developmental biology is well studied in this organism, research in this organism gives information regarding aging which is not possible with other animals.
6. *C. elegans* strains can be frozen and stored for long time without losing viability and they become alive when they are thawed. Therefore, the worms can be stored for a long period of time.
7. Developmental fate of all 959 cells of the *C. elegans* has been mapped.

3.6 Fruit fly: *Drosophila melanogaster*

Commonly called the fruit fly, *Drosophila melanogaster* is usually found around rotten banana and grapes. The fruit fly has an important place in genetics and genomics as it has been studied as a model organism since 1910. It was used by T. H. Morgan for his genetic studies and he developed the famous 'Flylab' at Columbia University. He made important discoveries on the arrangement of genes in chromosomes by constructing the first genetic map. As with other model organisms, this fly has a small structure, simple diet, large progenies and a short life cycle. The most important aspect of the fly genome is its salivary chromosome, which is large with several banding patterns that allow the genes to be mapped easily.

Thomas H. Morgan

T. H. Morgan was born on 15 September 1857 in the USA, and worked on *Drosophila* genetics and developed the famous 'Flylab' at Columbia University. He was awarded the Nobel Prize in 1933 for his pioneering work on chromosomal theory of inheritance.

The *Drosophila melanogaster* genome was sequenced in 2000 using the WGS strategy. The total genome size is 180 Mbp, of which the 120 Mbp euchromatic region has been sequenced. This is the first multicellular eukaryotic organism to be sequenced using the whole genome shotgun sequencing strategy. Surprisingly, the more complex fly had few genes higher than *C. elegans*, only 13,600. The *D. melanogaster* genome sequence is available in the **FlyBase** database. About 60 per cent of the genomic DNA is non-coding. The Y chromosomes contain mostly the heterochromatic region but they contain 16 male-related genes. About 75 per cent of the human disease genes are orthologous to *D. melanogaster* genes. Comparison of genomes of *D. melanogaster* and humans showed that many genes are conserved. Well-known examples are homeobox genes, nervous system development, etc.

Drosophila melanogaster

Kingdom	Animalia
Phylum	Arthropoda
Class	Insecta
Order	Diptera
Family	Drosophilidae
Genus	*Drosophila*
Species	*D. melanogaster*

Salient features of *Drosophila melanogaster*

1. Small size (2–3 mm length).
2. Easy to rear in the bottles on artificial diet.
3. Large fecundity (100 eggs per time).
4. Short life cycle (ten days at room temperature).
5. Sexual dimorphism.
6. Presence of salivary chromosome (polytene chromosome with puffs representing transcriptionally active region).

3.7 Plant genome

3.7.1 Mouse-ear cress: *Arabidopsis thaliana*

A. thaliana is a small flowering weed plant which belonging to the mustard family. It is analogous to *D. melanogaster* and *C. elegans* in the animal kingdom. It is used as a model organism to study plant biology. Studies on *Arabidopsis* started in the early 1900s, but only in 1998 was it declared a model organism. It has no direct relevance to agriculture as it is considered a weed. It is present across the world and there are many ecotypes or accessions available. For molecular biology and genome studies, cultivars such as Columbia and Landsberg are used. It has five pairs of chromosome (2n = 10) and genome size is 157 million base pairs. It has the smallest genome size among flowering plants. The complete genome sequence was published in 2000 and the number of predicted genes is 25,498, which are spread throughout the genome. The *Arabidopsis thaliana* Information Resource (TAIR) maintains the information pertaining to *A. thaliana*. One of the important aspects of transgenic development in *A. thaliana* is the floral dip method which avoids tissue culture. As many as 30,000 gene knockout collections are available for study.

Arabidopsis thaliana

Kingdom	Plantae
Division	Magnoliophyta
Class	Magnoliopsida
Order	Capparales
Family	Brassicaceae
Genus	Arabidopsis
Species	*A. thaliana (L.) Heynh*

Salient features of *Arabidopsis thaliana*

1. Small size plant (6–12 inch height).
2. Can be grown in Petri dish.
3. Life cycle (5–6 weeks).
4. Small genome compared to other plant genomes.
5. Large number of seeds per plant (10,000 per plant).
6. Easy to generate transgenic plants.
7. Genome contains much less repetitive DNA.
8. Well studied for light sensing and flower development.
9. Largest collections of mutants are available.
10. Translucent nature of the plant parts can be used to take fluorescent images to perform *in situ* analysis.

3.7.2 Rice genome: *Oryza sativa*

Rice (*Oryza sativa L.*) is one of the most important food crops in the world. Since rice is used by more than one-third of the world's population, further improvement is needed to meet future needs. Rice is cultivated in tropical countries of the world and is one of the earliest crops domesticated by man. A large number of studies on rice have been done and the biochemistry, physiology and genetics of rice have been well documented. Some 12,000 rice varieties are cultivated throughout the world. Therefore, as a reference rice variety, Nippon bare was selected for sequencing.

Rice is one of the commercially important crops, having a small genome. Rice has 12 chromosomes with a genome size of 430 Mbp and a C value of 0.43 to 0.45 pg. Rice genome sequencing was initiated in 1997. Japan has contributed about 55 per cent of the rice genome sequencing. Japan, the United States of America, China, Taiwan, Korea, India, Thailand, France, Brazil, and the United Kingdom participated in rice genome sequencing. To coordinate different agencies taking part in the rice genome sequencing project, the International Rice Genome Sequencing Project (IRGSP) under the guidance of Japan was started in 1997. The draft sequence of the rice genome sequence was announced in December 2002 and the rice genome sequence was completed in 2004. About 37,544 genes were predicted from the rice genome sequence. The rice genome sequence information and the genes identified in rice can be used to study other important food crops belonging to the cereal family, like maize and wheat, by comparative genomics.

Oryza sativa

Kingdom	Plantae
Division	Angiosperms
Class	Monocots
Order	Poales
Family	Poaceae
Genus	Oryza
Species	*O. sativa*

Salient features of *Oryza sativa*

1. Important food crop: over two-thirds of the world's population depend on this crop.

Published by Woodhead Publishing Limited

2. Small genome (430Mbp) among crop plants.
3. Well studied in terms of genetics, physiology.
4. Immediate use to a large section of people.

3.8 Animal genome

3.8.1 Puffer fish: *Fugu rubripes*

Fugu rubripes is popularly known as the puffer fish, and the literal meaning of *Fugu* is river pig. In Japan, a delicious dish is prepared with this fish. In 1989, Sydney Brenner suggested using it as a genome model to study vertebrate development and it is one of the first vertebrate genomes to be completely sequenced. The genome size is approximately 400 Mbp. The draft genome sequence was published in 2002. The total number of genes predicted was 28, 379, of which 17,504 are novel genes and only 1136 have been completely characterized. Since the *Fugu* genome shares considerable gene **synteny** with the human genome sequence, it can be used for annotation of the human genome with respect to vertebrate development.

Fugu rubripes

Kingdom	Animalia
Phylum	Chordata
Class	Actinopterygii
Order	Tetraodontiformes
Family	Tetraodontidae
Genus	Fugu
Species	*F. rubripes*

Salient features of *Fugu rubripes*

1. Has many genes and same regulatory elements present in the human genome; therefore it is used for human genome annotation.
2. Smallest vertebrate genome (390 Mbp).
3. *Fugu* is widely grown in farms in Japan.
4. It is a delicious dish prepared in Japan.
5. It contains a potent neurotoxin.

Published by Woodhead Publishing Limited

3.8.2 Mouse: *Mus musculus*

The mouse is considered one of the best model organisms for human healthcare. The mouse is an important biomedical model organism whose genome sequence is very closely related to human genome sequence. The mouse genome is 14 per cent smaller than the human genome but most genes are functionally homologous to their human counterparts. The order of genes in the chromosomes of humans and mice is conserved. From a higher degree of gene order conservation it is concluded that both species share a recent common ancestor. More than 99 per cent of the human genes have homologous genes in the mouse genome. It is the best animal model to study human diseases. Many of the diseases afflicting human beings are also detected in a mouse. Mutations that cause a particular disease in human are seen to cause the same disease in the mouse; therefore, it is used as a model organism to identify the genes responsible for human diseases.

The Mouse Genome Database (MGD) is a comprehensive storehouse of highly curated information pertaining to the physical and genetic maps with consensus positions and gene products of mouse genome sequence. The mouse has been used for testing of the drugs for many decades, and also many genes responsible for cancer and other human diseases were identified. It is possible to create gene knock-out in a mouse to study their functions and to create disease conditions for study. The Mouse Genome Sequencing Consortium (MGSC) was developed to organize and coordinate the Mouse Genome Project. The major institutes involved in mouse genome sequencing are the Whitehead/Massachusetts Institute of Technology, the Centre for Genome Research, the University of Washington, the Wellcome Trust, and the Sanger Institute.

Mus musculus

Kingdom	Animalia
Phylum	Chordaatum
Class	Mammalia
Order	Rodentia
Family	Muridae
Genus	Mus
Species	*Mus musculus*

Salient features of *Mus musculus*

1. It is a mammal.
2. A variety of inbred strains are available.
3. A transgenic mouse can be created easily.
4. Random as well as directed mutagenesis is possible.
5. A huge pool of genetic data is available.
6. Size of the mouse is small when compared to the other animals, adding to ease of handling.
7. Short life period (2 months).
8. Large number of offspring (10–12 per time).
9. New mutant strains can easily be introduced by irradiation and using other chemical mutagens.
10. Foreign DNA can be easily introduced through embryo manipulation methods.
11. Interspecies hybridization is possible because of availability of more than one species like M. *musculus* and M. *spretus*, which is helpful in genetic mapping.
12. Transgenic technology is highly advanced. Specific genes can be knocked out by homologous recombination.

3.8.3 Rat: *Rattus rattus*

Rattus rattus is an important experimental animal. The rat genome (2.75 Gbp) is smaller than the human genome (2.9 Gbp) and little larger than the mouse genome (2.6 Gbp), but all three species have a similar number of genes. A comparison of these three genome sequences revealed that the gene order is conserved. It has also been identified that almost all human genes associated with diseases have orthologues in rat, which makes it an animal of choice as animal model for testing drug action. Rats are carriers of world-threatening human diseases like cholera, plague, leptospirosis and other diseases.

Rattus rattus

Kingdom	Animalia
Phylum	Chordata
Class	Mammalia
Order	Rodentia

Published by Woodhead Publishing Limited

Family	Muridae
Genus	Rattus
Species	*R. rattus*

Salient features of *Rattus rattus*

1. An important model for toxicology studies.
2. Human orthologous genes are also identified in rat genome.
3. Many human diseases can be induced in the rat.

3.9 The Microbial Genome Project

Table 3.1 lists the model organisms and their genomic features used in the Microbial Genome Project.

Table 3.1   List of model organisms and their genomic features

Sl. No.	Scientific name	Genome size (kbp)	Strategy	No. of genes predicted
1	Lambda	48	Whole genome shotgun	
2	*Haemophilus influenzae*	1830	Whole genome shotgun	1743
3	*E. coli*	4600	Whole genome shotgun	4000
4	*Saccaharomyces cerenisiae*	12,156	Hole genome shotgun	6275
5	*Cenorhabdidtis elegans*	97,000	Clone-by-clone	10,500
6	*Arabidopsis thaliana*	1,55,000	Clone-by-clone	24,488
7	*Drosophila melanogaster*	1,60,000	Whole genome shotgun	14,000
8	*Oryza sativa*	4,30,000	Clone-by-clone	37,544
9	*Fugu rubripes*	3,65,000	Whole genome shotgun	28,379
10	*Mus musculus*	26,00,000	Hybrid	30,000
11	*Rattus rattus*	27,50,000	Clone-by-clone	30,000

Review questions and answers

1. *What is a model organism?*

An organism whose genetics and physiology are studied in detail, and is amenable for scientific research. The outcome of the study on this organism is useful to understand the function of other organisms, specifically humans.

Published by Woodhead Publishing Limited

2. *Why do you have to use a model organism for genome sequencing?*

The model organism used for genome sequencing is called a genetic model. It has comparatively more information about the gene.

3. *What was the first multicellular organism whose genome was completely sequenced?*

Caenorhapditis elegans.

4. *What is the name of the first free-living organism whose genome was completely sequenced?*

Haemophilus influenzae.

5. *Suggest a model organism which can be studied for cell division.*

Saccharomyces cerevisiae.

6. *Name a model organism which is suitable for developmental biological studies.*

C. elegans or *Drosophila melanogaster.*

7. *Arabidopsis is a weed plant, but it was the first plant genome sequenced. Why is it considered a model organism for plant biology?*

It has the smallest genome in the plant kingdom.

8. *List the essential features of model organisms for genomics study.*

- Cost of maintenance should be low.
- Ethical problems should not arise.
- Share certain degree of relationship with human genome and physiology.
- A particular biological phenomenon is easier to understand in that organism rather than in humans.
- Easy genetic manipulation.
- Availability of genetic information such as the genetic map, the physical map and markers for mapping.
- Short life cycle.

- High fecundity.
- Information derived from the model organism should have applications in human health care.

Recommended reading

Adams, M.D. et al. (2000) 'The genome sequence of *Drosophila melanogaster*', *Science*, 287: 2185–95.

Blake, J.A., Eppig, J.T., Bult, C.J., Kadin, J., Richardson, A. and Mouse Genome Database Group (2005) 'The mouse genome database (MGD) update and enhancements', *Nucleic Acid Research*, 34(suppl.): D562–D567.

Blattner, F.R. et al. (1997) 'The complete genome sequence of *Escherichia coli* K-12', *Science*, 277: 1453–74.

Borkovich, K.A. et al. (2004) 'Lessons from the genome sequence of *Neurospora crassa*: tracing the path from blueprint to multicellular organism', *Microbiology and Molecular Biology Review*, 68(1): 1–108.

Casjens, S. (1998) 'The diverse and dynamic structure of bacterial genomes', *Annual Review of Genetics*, 32: 339–77.

Fleischmann, R.D., Adams, M.D., White, O., Clayton, R.A., Kirkness, E.F., Kerlavage, A.R., Bult, C.J., Tomb, J.F., Drougherty, B.A., Merrick, J.M. et al. (1995) 'Whole-genome random sequencing and assembly of *Haemophilus influenzae* Rd.', *Science*, 28: 496–512.

Goffeau, A. et al. (1996) 'Life with 6000 genes', *Science*, 274(546): 563–7.

International Rice Genome Sequencing Project (2005) 'The map-based sequence of the rice genome', *Nature*, 436: 793–800.

Mouse Genome Sequence Consortium (2002) 'Initial sequencing and comparative analysis of the mouse genome', *Nature*, 420: 520–62.

Rat Genome Sequencing Project Consortium (2004) 'Genome sequence of the brown Norway rat yields insights into mammalian evolution', *Nature*, 428: 493–521.

Sanger, F. et al. (1982) 'Nucleotide sequence of bacteriophage DNA', *Journal of Molecular Biology*, 162: 729–73.

Stothard, P. et al. (2005) 'BacMAp: an iterative picture atlas of annotated bacterial genomes', *Nucleic Acid Research*, 33: D317–D320.

The Arabidopsis Genome Initiative (2000) 'Analysis of the genome sequence of the flowering plant *Arabidopsis thaliana*', *Nature*, 408: 796–815.

The *C. elegans* Sequencing Consortium (1998) 'Genome sequence of the nematode *C. elegans*: a platform for investigating biology', *Science*, 282: 2012–18.

Web addresses

http://www.arabidopsis.org/.
http://www.broadinstitute.org/annotation/genome/neurospora/MultiHome.html.
http://www.ebi.ac.uk/genomes/bacteria.html.

http://www.ensembl.org/Mus_musculus/.
http://www.genome.wisc.edu/.
http://microbialgenomics.enegry.gov/.
http://rgp.dna.affrc.go.jp/E/index.html.
http://www.wormbook.org/.
http://www.yeastgenome.org/.

Published by Woodhead Publishing Limited

4

High capacity vectors

Abstract: Before genome sequencing, genomic DNA is fragmented and cloned into vectors. Then the recombinant DNA is introduced into a host and is maintained for further use. The group of clones representing the entire genome is known as the genomic library. High capacity cloning vectors are used in the construction of the genomic libraries for genome sequencing. In this chapter different high capacity cloning vectors, such as cosmid, fosmid, PAC, BAC and YAC, are discussed with respect to their essential features, advantages and disadvantages.

Key words: BAC, cosmid vector, fosmid vector, high capacity vectors, PAC, YAC.

Key concepts

- High capacity vectors are used to construct genome libraries.
- The cloning capacity of high capacity vectors is more than that of plasmid vectors.
- Cloning in high capacity vectors reduces the number of clones to be handled for each genome.
- Fosmid libraries are specifically used for gap closing as they are capable of maintaining difficult DNA regions like repetitive sequences.
- BAC vectors have high capacity up to 300 kbp, and are used to maintain the foreign DNA stably for many generations.
- YAC vectors have highest cloning capacity (up to 3000 kbp) but they do not maintain the foreign DNA stably.

4.1 Introduction

When large-scale genome projects were undertaken in the 1990s, one of the objectives of the genome sequencing projects was to develop **high capacity**

Table 4.1 High capacity vectors and their properties

Sl. No.	Name of the vector	Cloning capacity (kbp)	Method of introduction into the host	Host
1	Cosmid vector	40–45	transduction	*E. coli*
2	Fosmid vector	100		*E. coli*
2	P1 derived vector	70–100	transduction	*E. coli*
3	PAC	130–150	electroporation	*E. coli*
4	BAC	120–300	electroporation	*E. coli*
5	YAC	150–3000	transformation	*S. cerevisiae*

vectors. *Cloning vectors* were developed in 1970s. Initially, the developed vectors had a cloning capacity of less than 10 kbp. They were useful for genomic library construction and cDNA library construction. High capacity vectors were developed in order to reduce the number of clones that needed to be handled. Further improvements were also made to increase the cloning capacity of the vectors. Large capacity vectors are used for physical mapping of the genome, clone **contig** preparation and to maintain the genome to be sequenced in a library. Many such vectors were developed and they are called high capacity vectors. The cloning capacity of these vectors is above 40 kbp (Table 4.1).

The first step in genome sequencing project is to construct a genome library of that organism. The genome library serves three purposes, namely, separation of DNA from the rest of the genomic DNA fragments for sequencing, amplification of a particular DNA fragment, and the maintenance of genomic DNA fragments for many generations. The clone libraries also play a vital role in physical mapping and genome sequence assembly.

4.2 Cosmid vectors

Cosmids are medium-sized cloning vectors. The cloning capacity of these vectors is 35–45 kbp. The first cosmid vector was described by Collins in 1978. Cosmid vectors are developed by combining the features of the **plasmid** vector and the bacteriophage vector. Origin of replication, *multiple cloning* site and selectable marker are obtained from the plasmid and only the cohesive site or **cos site** region is taken from lambda phage. These are fused together to obtain the cosmid vector. Approximately 200 bp lambda

Published by Woodhead Publishing Limited

phage sequence is cloned into the cosmid vector. This consists of cosN, cosB and cosQ. A cosmid vector may have one or two cos sites. Cosmid vectors are used in the construction of genomic libraries. The cloning of a foreign DNA in cosmid vector involves the following steps: (1) ligation of the foreign DNA between two cos sites; (2) making a concatemeric DNA; (3) *in vitro* **packaging** to introduce the DNA into the phage head to form the matured phage particle; and (4) introduction of the cloned DNA into *E. coli* by transduction. After their entry into the host cell, the cosmids are maintained as plasmids. Usually cosmids are maintained as high copy plasmids. Cosmid vectors can also be used as a shuttle vector by incorporating SV40 origin of replication.

Essential features of a cosmid vector

1. *E. coli*-based cloning vector.
2. Constructed by combining plasmid origin of replication and selectable marker gene and bacteriophage (cos site) for genetic elements.
3. Cloning capacity up to 45 kbp.
4. Used for constructing genomic library.

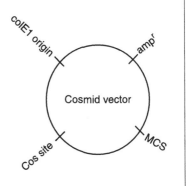

4.3 Fosmid vectors

The need for an alternative high capacity cloning vector was felt when the genomic library constructed using cosmid vectors was unstable due to the high copy of the cosmid vector. Therefore, a new cloning vector with a low copy number was developed based on F-factor plasmid and this is called the **fosmid** vector. Fosmid is a cosmid cloning system but it has an **F-factor** origin of replication to control the copy number of the vector. Fosmids are used for genomic library construction of complex genomes. A fosmid is a low copy number plasmid constructed by combining the features of F-factor plasmid and the cos site. These libraries are stably maintained as they are low in copy number, i.e. one copy per cell. The fosmid vector library is used to check the quality of the sequence generated from genome projects. pFOS1 was the first developed fosmid vector. The main application of a fosmid vector is that a stable genomic library can be constructed and maintained for longer. Fosmid libraries are useful in filling gaps in the contig assembly. They are also used for detecting deletions, insertion and rearrangements, etc. The size of the fosmid vector is ~8 kbp. It has parB and parA genes to maintain the copy number of the fosmid. A kanamycin-resistant gene or an ampicillin-resistant gene is used as a selectable marker.

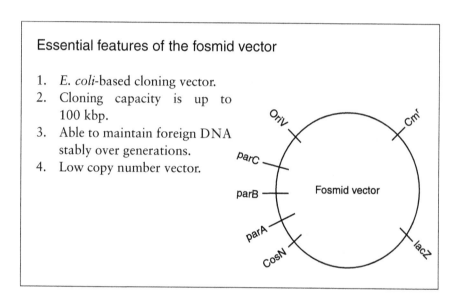

Essential features of the fosmid vector

1. *E. coli*-based cloning vector.
2. Cloning capacity is up to 100 kbp.
3. Able to maintain foreign DNA stably over generations.
4. Low copy number vector.

4.4 Bacteriophage P1 derived vector

This is a high capacity cloning vector based on P1 bacteriophage. It has a *cis* acting element derived from a bacteriophage to accommodate the foreign DNA up to 70 kbp–100 kbp. After cloning in the P1 derived vector, the recombinant P1 derived vector is *in vitro* packaged using the sequence pac and P1 lysate. It has two lox P sites. After injecting into *E. coli*, recombinase enzyme present in *E. coli* cells circularizes the recombinant P1 derived vector by combing lox P sites. It is maintained as one copy per cell. It has a positive selection marker SacB. It has an additional P1 lytic replication and it is under the control of the lac promoter.

Essential features of a P1 derived vector

1. Cloning capacity is 75 kbp to 100 kbp.
2. Maintains one copy per cell.
3. Foreign DNA is maintained stably.

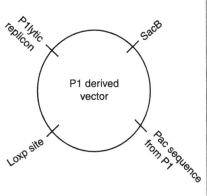

4.5 P1 derived artificial chromosome (PAC)

PACs are constructed by combining the best features of P1 bacteriophage vector and BAC. No chimeras or instability is associated with this vector and it is used for genome mapping. It is introduced by **electroporation**. The carrying capacity of the PAC is 60 kbp–150 kbp.

Essential features of a PAC vector

1. *E. coli*-based vector.
2. Low copy number.
3. Stable maintenance of foreign DNA.
4. Cloning capacity is up to 150 kbp.

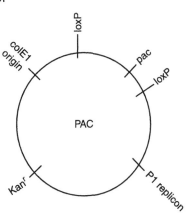

4.6 Bacterial artificial chromosomes (BAC)

This is a high capacity cloning vector developed for genome sequencing projects. It was developed to prepare a large insert clone library for genome sequencing projects. The cloning capacity of **BAC** vectors is up to 300 kbp. It gained popularity in genome sequencing projects due to its high cloning capacity. It is maintained as a single copy plasmid. It is characterized by stable maintenance even after more than a hundred generations. Its origin of replication is F plasmid. It has three genes for partitioning such as parA, parB and parC. The host bacteria should be deficient in homologous recombination (it should be recA⁻). The size of the BAC vector is 7.4 kbp. After ligating with the foreign DNA, recombinant BAC is introduced into the *E. coli* host by **electroporation.**

Essential features of a BAC vector

1. *E. coli*-based highest capacity cloning vector.
2. Single copy plasmid.
3. Most widely used vector for genome sequencing projects.
4. Constructed based on F factor plasmid.
5. Introduced into the host by electroporation.

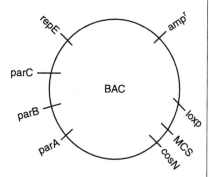

4.7 Yeast artificial chromosome (YAC)

Cloning capacity is up to 3000 kbp and is used for physical mapping. It has telomere, centromere, origin of replication and selectable makers. After cloning the recombinant **YAC** vector, it is maintained in the cell as a linear DNA-like chromosome, as the name suggests. CEN sequence obtained from the yeast plays a major role in maintaining the copy number of the vector after cell division. **Autonomously replicating sequences (ARS)** are equal to the **origin of replication** to the plasmid vector as it helps to initiate replication. The commonly used selectable markers are URA3 and Sup4. Partially digested genomic DNA yields large fragments that are cloned into the YAC vector arms.

Essential features of a YAC vector

1. Yeast-based cloning vector.
2. Possesses highest cloning capacity.
3. Maintained as linear DNA-like chromosome.
4. Introduced into the yeast cells by electroporation.

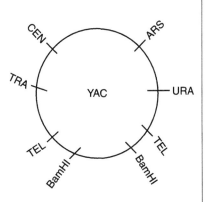

Review questions and answers

1. *What is the significance of high capacity vectors for genome mapping?*

Cloning in high capacity vectors reduces the number of clones to be handled for each genome.

2. *Fosmid vectors are used to construct the stable genomic clone libraries in genome projects. Justify the statement.*

Fosmid is a low copy number vector constructed by combining the features of F plasmid, and cos site and it also has parA and par B genes. These two features help to maintain the foreign DNA stable in the host cell.

3. *YAC can carry larger inserts than any other high capacity vectors, but they are not used for genome projects. Validate the statement.*

Because foreign DNA cloned in YAC vectors undergoes *in vivo* recombination, it is not maintained properly.

4. *Discuss the importance of BAC in the context of genome sequencing.*

The BAC vector is mostly used for genomic library construction as high capacity vector to take up foreign DNA and also maintain it stably over generations.

Published by Woodhead Publishing Limited

Recommended reading

Burke, D.T., Carle, G.F., and Olson, M.V. (1987) 'Cloning of large segments of exogenous DNA into yeast by means of artificial chromosome vectors', *Science*, 236: 806–12.

Ioannou, P.A., Amemiya, C.T., Garnes, J., Kroisel, P.M., Shizuya Chen, C., Batzer, M.A. and de Jong, P.J. (1994) 'A new bacteriophage P1 derived vector for the propagation of large human DNA fragments', *Nature Genetic*, 6: 84–9.

Kim, U.J., Shizuya, H., Jong, J.P., Birren, B. and Simon, M.I. (1992) 'Stable propagation of cosmid sized human DNA insert in an F-factor based vector', *Nucleic Acid Research*, 20: 1083–5.

Shizuya, H., Birren, B., Kim, U.J., Mnacino, U., Slepak, T., Tachiiri, Y. and Simon, M. (1992) 'Cloning and stable maintenance of 300-kilobase-pair fragments of human DNA in *Escherichia coli* using an F factor-based vector', *Proceedings of National Academy of Sciences USA*, 89: 8794–97.

Strenberg, N. (1990) 'Bacteriophage P1 cloning system for the isolation, amplification and recovery of DNA fragments as large as 100 kilobase pairs', *Proceedings of National Academy of Sciences USA*, 87: 103–7.

Web address

http://www.ornl.gov/sci/techresources/Human_Genome/elsi/cloning.shtml.

5

DNA sequencing methods

If you want to practice biology, do it on the leading edge and if you want to be on the leading edge, invent new tools for deciphering biological information.

William Dreyer

Abstract: Two methods for DNA sequencing were developed at the same time to reveal the order of nucleotides in a DNA fragment. The chemical degradation method of DNA sequencing involves base-specific cleavage of the DNA strands while the dideoxy chain termination method works by enzymatic termination of the growing strands. The dideoxy chain termination method can be carried out under mild conditions using enzymes and does not involve hazardous chemicals whereas the chemical degradation method involves toxic chemicals. Many advances were introduced in the dideoxy chain termination method to increase the speed and the accuracy of sequencing. Therefore, this method was widely used in high throughput genome sequencing technology. Recently, the next generation of sequencing methods has been developed which are not based on the dideoxy chain termination principle and these methods are also discussed in this chapter.

Key words: base calling, capillary electrophoresis, DNA sequencing method, dNTP, ddNTP, electrophoresis, fluorescent labelling, radioactive labelling, sequencing gel.

Key concepts

- Determining the order or arrangement of nucleotides in a DNA fragment is termed DNA sequencing.
- Methods for DNA sequencing are the Maxam and Gilbert method (chemical degradation method) and the Sanger's method (chain termination method).

- The chemical degradation method involves base-specific cleavage of the template DNA to be sequenced into nested set of DNA fragments in four base-specific reactions.
- The nested sets of fragments are separated on sequencing gel which resolves the DNA fragments of one nucleotide difference.
- Radioactive phosphorus (^{32}P or ^{33}P) is added to the template or primer before the sequencing reaction in order to locate the fragments after separation.
- Autoradiography is used to locate the DNA fragments as bands on X-ray sheet. From the relative banding position, the nucleotide sequence is deduced.
- As radioactivity is hazardous to human health, non-radioactive labelling methods have also been developed.
- The chemical cleavage method is not as popular as the dideoxy chain termination method because of the toxic chemicals used in this method.
- The chain termination method is amenable for automation and is being used for genome sequencing.

5.1 The history of DNA sequencing

Basic information required for any cellular process is contained in the genetic material, which is made up of long strands of deoxyribonucleic acids. The order of bases in the polynucleotide strand determines the function of the gene products. Therefore, it is important to know the sequence of a gene or the genome. Sequencing of DNA fragments has resulted in the chemical identification of genes and genomes. Nowadays, DNA sequencing has become an integral part of molecular biology and genetic engineering experiments.

The history of nucleic acid sequencing dates back to the 1960s. It all started with the sequencing of ribonucleic acid (**RNA**). The RNA sequencing method was completely dependent on enzymes. In 1963, Robert Holley successfully sequenced the alanine tRNA of yeast. The total number of bases deduced was only 77 but this took a very long time. The column-purified RNA samples were used for sequencing. Overlapping RNA fragments were obtained after treatment with **endonuclease** digestion. The digested fragments were subjected to **exonuclease** to yield individual nucleotides from one end. The resulting nucleotides were identified by thin-layer chromatography (TLC).

A similar procedure was tried for DNA with some modifications but it was not successful in yielding the DNA sequence. This was due to the fact that no base-specific DNases were available, and also, the fact that different bases of DNA are chemically similar. In 1973, Gilbert and Maxam sequenced DNA after transcribing it into RNA. The plus–minus method of DNA sequencing was developed by Sanger in 1975. Continuous hard work by Maxam and Gilbert at Harvard University, USA, and Frederick Sanger at Cambridge University, UK, paved the way for the development of modern DNA sequencing methods in the same year (1977).

Milestones in nucleotide sequencing

1963	Robert Holley sequenced the alanine tRNA of yeast.
1972	Walter Fiers sequenced MS2 bacteriophage using RNA sequencing.
1973	Walter Gilbert and Allan Maxam with wandering spot analysis sequenced lac operator.
1975	Sanger and Coulson developed the plus–minus sequencing of ΦX174 bacteriophage.
1977	Maxam and Gilbert discovered the chemical cleavage method of DNA sequencing.
1977	Frederick Sanger discovered the chain termination method of sequencing.
1980	Sanger and Gilbert won the Nobel Prize for DNA sequencing.
1986	Leroy E. Hood invented the semi-automated DNA sequencing machine.
1997	Applied Biosystems Model ABI370, fully automatic sequencing machine with capillary electrophoresis.
2005	Genome Sequencer GS20, 454 Life Sciences, Roche.
2006	Genome Analyser, Solexa/Illumina.
2007	SOLID Applied Biosystem.

5.2 Steps in DNA sequencing

The first generation sequencing methods (the **dideoxy chain termination method** and the chemical degradation method) involve similar steps. They differ in the principle by which sequencing reactions are carried out (see Figure 5.1). In both methods, a nested set of labelled DNA fragments is prepared and subjected to **chemical cleavage** or enzymatic reactions. The sequencing reactions generate DNA fragments that differ from each other by one nucleotide. These DNA fragments are separated on a **sequencing gel** (thin gel made of polyacrylamide) under very high voltage and their bands are viewed on an

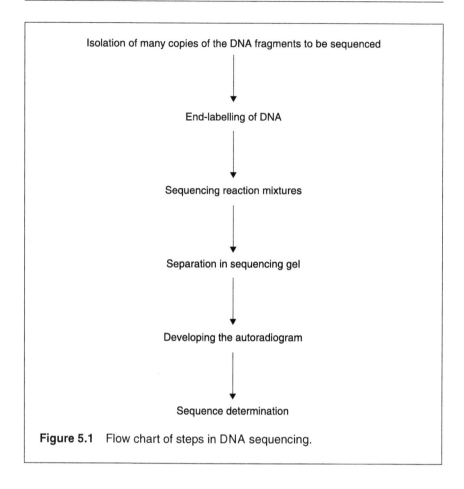

Isolation of many copies of the DNA fragments to be sequenced

End-labelling of DNA

Sequencing reaction mixtures

Separation in sequencing gel

Developing the autoradiogram

Sequence determination

Figure 5.1   Flow chart of steps in DNA sequencing.

X-ray sheet. From the pattern of the DNA bands, the corresponding bases can be deduced. From a single sequencing reaction, the number of bases that could be determined ranged between 300 and 400. This is called the read length.

5.3 Chemical degradation method of DNA sequencing

This method is also known as the Maxam and Gilbert method of DNA sequencing named after its discoverers in 1973. Originally, this method was devised to sequence a small DNA fragment of 24 bases and this technique was known as 'wandering spot analysis', later, the technique was further improved and used for sequencing large DNA fragments (Maxam and Gilbert, 1977). Walter Gilbert shared the Nobel Prize for DNA sequencing with Frederic Sanger in 1980.

The principle involved in this method is chemical modification of specific bases followed by cleavage of the DNA strand at that place. This occurs in two steps, in the first step, base-specific reagents and the reaction condition are set to modify/remove a base from the DNA strand, leaving the sugar phosphate backbone intact. In the second step, the base-removed DNA strand is made sensitive to phosphodiester cleavage by the chemical piperidine. The strand breakage results in a set of DNA fragments whose lengths differ by one nucleotide. They can be separated on denaturing polyacrylamide gel (sequencing gel) and identified using an **autoradiogram** (Figure 5.2).

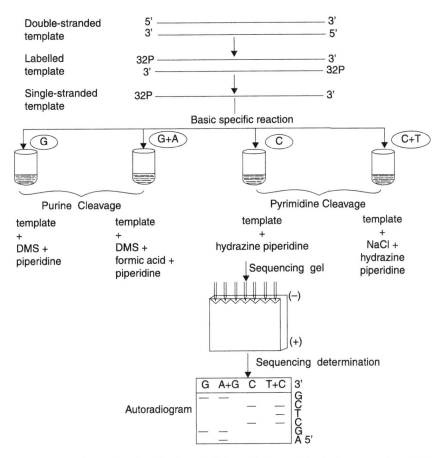

Figure 5.2 Steps involved in chemical degradation method of sequencing. DNA to be sequenced is purified and labelled with ^{32}P at 5' end. The single-stranded labelled DNA template is separated and used for base specific reactions. Four base-specific reactions are set up to cleave the template at G, G+A, C, C+T. These reactions are stopped and run in sequencing gel and the DNA sequence is determined from the autoradiogram.

1. **End-labelling** of DNA fragments: DNA fragments to be sequenced are purified through column chromatography and used for end-labelling. A radioactive isotope of phosphorus (^{32}P or ^{33}P) is incorporated into the 5′ end of the purified DNA fragments through the enzymatic method. DNA-modifying enzymes such as phosphatases and kinases remove or add the phosphate group to the DNA fragments, respectively. The addition of radioactive phosphate allows it to identify extremely low amounts of DNA (ng) in the gel which is otherwise invisible to the naked eye. Treatment of DNA fragments with the alkaline phosphatase enzyme results in the removal of the phosphate group from the 5' end of the DNA. The resulting dephosphorylated DNA fragments are then treated with polynucleotide kinase (PNK) in the presence of radioactive deoxyadenosine triphosphate (γ-^{32}P dATP). PNK adds a phosphate group to the 5' end of the DNA fragments. Uniformly end-labelled DNA fragments are thus generated for further reaction.

2. Chemical modification and cleavage: 5'-end-labelled DNA fragments are used for chemical cleavage reactions. Four separate reactions are carried out simultaneously in four tubes marked as G, G+A, C, and C+T. Specific chemicals are added to each tube and conditions are adjusted for the reaction. In all four reactions, the following two steps are common: (i) base modification and removal of a base directly from the DNA strand; and (ii) phosphodiester cleavage of the DNA strands where a base is removed (Table 5.1).

3. Separation of DNA fragments on sequencing gel: The DNA fragments generated have a common starting point with a radioactive phosphate group at the 5' end. Denaturing polyacrylamide gel is used to separate the DNA fragments. A long (100 mm) and thin (0.1–0.5 mm) polyacrylamide

Table 5.1 Base-specific modification and cleavage of DNA in chemical cleavage reactions

Base modified and/removed	Type of modification	Chemicals involved	Reaction condition
Guanine	Methylation at N_7	Dimethylsulfate	Piperidine at 90°C for 30 min.
Guanine and Adenine	Depurination	Dimethylsulfate + 1.5 M formic acid	Piperidine at 90°C for 30 min.
Thymine and Cytosine	Base ring opening	Hydrazine	Piperidine at 90°C for 30 min.
Cytosine	Base ring opening	Hydrazine + 1.5 M NaCl	Piperidine at 90°C for 30 min.

Published by Woodhead Publishing Limited

gel is prepared with 8 M urea. Urea is added to the gel in order to remove any secondary structure formed by single stranded DNA as this affects their electrophoretic mobility and band position. The wells are formed using shark's tooth comb which holds less volume of the sample. The four reactions mixtures are loaded in the sequencing gel side by side. The sample is loaded with automatic pipettes fitted with a flat microcapillary tip that fits the shark's tooth well. A 30 gauge needle is used for loading and the needle is washed thoroughly with 0.5X TBE buffer before each loading. Usually the order of loading is TCGA (this is the preferred order as the C and G lanes create a gel compression problem) or CTAG (preferred by Sanger, pyrimidines followed by purines according to their alphabetical order). A high voltage is applied (~2100 V for 40 cm gel) which is required to maintain the gel temperature at 45°C to 50°C. The sequencing gel can resolve DNA fragments of size 15–400 nucleotide. When the bromophenol blue dye reaches the bottom of the gel, the electric current supply is discontinued in order to stop electrophoresis.

4. Developing an autoradiogram: In order to detect DNA bands, the gel is exposed to an X-ray film. After sequencing, the gel is kept in firm contact with the X-ray film by keeping them together inside a cassette. The radioactive phosphorus (^{32}P) present in the DNA fragments emits β rays with an energy of 1.7 MeV. This converts the silver halide present in the X-ray emulsion to silver ions. Upon developing the film, dark bands appear on the X-ray film which indicates the position of DNA fragments. This process is known as **autoradiography** and the X-ray film with bands is known as an autoradiogram.

5. Sequence determination: During electrophoresis, the movement of DNA fragments depend on their size (the shortest fragment will reach the bottom faster than the others). When the DNA fragments are cleaved, the other half of the DNA fragment will not be detected in the autoradiogram as they are not radioactive. Based on the banding position in the four base-specific reactions which are loaded side by side, the sequence is determined.

Advantages of the chemical degradation method of DNA sequencing

1. Template DNA fragment can be used for sequencing directly.
2. It is useful in studying methylation of DNA and genetic imprinting, but it does have many disadvantages.

Disadvantages of the chemical degradation method of DNA sequencing

1. The chemicals used are hazardous.
2. It is time consuming and labour intensive.
3. Not amenable to automation.
4. Only a small number of nucleotides can be read from a single reaction.

5.4 The chain termination method of DNA sequencing

This is an enzymatic method of DNA sequencing, first described by Sanger in 1977. A complementary DNA strand is synthesized by *DNA polymerase* in the presence of the dNTP mix, labelled **primer** or a labelled ddNTP. The principle of dideoxy chain termination method lies in the ability of the DNA polymerase to incorporate both the dNTP and ddNTP in the growing chain of the DNA synthesis. When dNTP is incorporated, the growing chain continues to grow further, in other words, DNA polymerase increases the nucleotides further. Whenever a ddNTP is added, further addition of either dNTP or ddNTP is prevented due to the absence of free 3'-OH group. This results in the termination of the growing chain of the DNA, hence the name chain termination. The nested set of chain-terminated strands has a common starting point with radio-labelled phosphorus (^{32}P), and the fragments can be identified on an autoradiogram which is developed after separation on a sequencing gel (Figure 5.3).

Frederick Sanger

Sanger was born on 13 August 1918 in England. Basically a protein chemist, he determined the sequence of insulin polypeptide and was awarded his first Nobel Prize in 1958 for having proved that proteins have a defined chemical structure. Later, he worked towards sequencing of RNA. The first RNA sequencing was reported by R. Holley before Sanger's group finished the task; he then turned his attention to the sequencing of DNA and developed the popular dideoxy chain termination method. He shared the Nobel Prize with Maxam and Paul

Berg in physiology or medicine in 1980. The Wellcome Trust and the Medical Research Council established a world-renowned sequencing centre named after him in 1992.

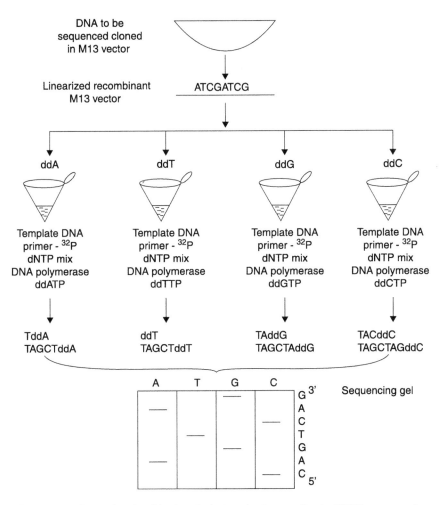

Figure 5.3 Steps involved in the chain termination method of DNA sequencing. DNA to be sequenced is cloned in the M_{13} vector and amplified in the *E. coli* host. Single-stranded DNA template is linearized and used for base-specific reactions. Four base-specific reactions are set up to synthesize complementary strands labelled primer, linearized template, DNA polymerase, dNTP and any one of the ddNTP. The reaction mixtures are run on a sequencing gel. Sequence is determined from the autoradiogram.

5.4.1 Steps involved in the chain termination method

1. Isolation and purification of template DNA. The dideoxy chain termination method requires a single-stranded template DNA. It is obtained by cloning the DNA fragment to be sequenced in an M_{13} cloning vector. M_{13} is a bacteriophage-based vector. When it multiplies inside the host bacteria, the progenies continue to be secreted into the media as viral particles containing a single-stranded viral genome. The recombinant M_{13} vector helps to obtain the single-stranded DNA for sequencing.
2. Preparation of a radio-labelled primer: The chain termination method requires the template to be copied by the DNA polymerase. It requires a primer for initiation. Universal primers based on M_{13} vector sequences are used for this purpose. Therefore, in the chain termination method, the primers are labelled using radioactive phosphate (^{32}P).
3. Setting up of sequencing reaction: Enzymatic synthesis of complementary strand with the following components in four separate tubes is carried out. Each tube contains common reaction components except for ddNTP. The common reaction components are template DNA, radio-labelled primer, dNTP mix, DNA polymerase. The unique component in each tube is the ddNTP. Each tube is marked either ddA, ddT, ddC, or ddG, corresponding to the respective ddNTP added. The tubes are incubated for a specified period of time for the reaction to proceed.
4. Separation of reaction mixture in sequencing PAGE gel: Same as that of the Maxam and Gilbert method.
5. Autoradiogram: Same as that of the Maxam and Gilbert method.
6. Deducing template sequence: The lanes are marked as ddA, ddT, ddC, and ddG, according to the type of ddNTP used. The sequence is determined from the bottom to the top of the autoradiogram. Since the sequence deduced from the autoradiogram is the complementary strand of the template being used for sequencing, the sequence of the original strand should be identified from the complementary base pairing to the determined sequence.

Applications of DNA sequencing

1. Characterization of genes and genomes.
2. Site-directed mutagenesis studies.

Published by Woodhead Publishing Limited

3. To locate protein–DNA interaction.
4. To construct phylogenetic trees based on multiple alignment of DNA sequence.
5. Recombinant DNA construction.

5.5 Advances in DNA sequencing methods

Large-scale DNA sequencing was possible only after the invention of automated DNA sequencing machines. Sanger's chain termination method was found to be more amenable to automation than the chemical degradation method (Table 5.2). The first improvement made by replacing the radio-labelling of DNA fragments with those of fluorescent dyes was introduced by Leroy Hood and his colleagues in 1985. This resulted in the development of the semi-automated DNA sequencer which was capable of performing automated **base calling**. An important aspect of automated DNA sequencing is that it uses fluorescent labelling as opposed to radioactive labelling. This completely eliminates radioactivity hazards and manual base determination. Later, many modifications were introduced in terms of instrumentation, sequencing chemistry, sequencing enzymes, etc. Automated DNA sequencing machines which are capable of performing sequencing with large capacities have been developed and marketed by many companies. The first automated DNA sequencer was marketed by Applied Biosystems. Automated sequencing development is the one of the factors that led to large-scale sequencing. This technique was 100 times faster than traditional DNA sequencing methods. This also reduced the cost of sequencing 100-fold from US$10 per finished base to US$1.

Leroy Hood

Hood was born on 10 October 1938 in the USA. He received his PhD in Biochemistry. He developed an automated DNA sequencer which played a major role in genome sequencing. He is one of the eleven scientists who started the human genome project. Hood also developed the protein synthesizer. Currently, he is working on high speed bar-coding techniques.

Published by Woodhead Publishing Limited

Table 5.2 Comparison of chemical cleavage method and chain termination method of DNA sequencing

Sl. No.	Chemical cleavage method	Chain termination method
	Differences	
1	Time required to set up reaction is more than one day	Only one or two hours
2	Sequence read per gel is approximately 150 bases	Sequence read per gel is approximately 300 bases
3	Template DNA can be used for sequencing reaction directly	Template DNA should be cloned for sequencing reaction
4.	Radiolabelling of template is involved	Radiolabelling of primer is involved
5	Only one way of introducing radiolabel.	Two ways of introducing radiolabel
6	DMS, hydrazine and piperidine are hazardous chemicals	No hazardous chemicals involved
7	Difficult to automate	Automation is possible
8	Not used for large-scale sequencing	Used for large-scale genome sequencing
9	Template strand is sequenced	Complementary strand is sequenced
	Similarities	
1	Nested set DNA fragments are generated at each end with a particular base	Nested set DNA fragments are generated at each end with a particular base
2	DNA fragments with one nucleotide difference are resolved on denaturing PAGE gel	DNA fragments with one nucleotide difference are resolved on denaturing PAGE gel
3	DNA fragments with common starting point (5') are generated	DNA fragments with common starting point (5') are generated
4	Sequence is read from 5' to 3' direction	Sequence is read from 5' to 3' direction

5.5.1 Fluorescent dye-labelled terminator

In the modern DNA sequencing technique, DNA fragments are labelled with fluorescent dyes rather than radioactive phosphate (^{32}P). Four different fluorescent dyes are covalently attached to each of the ddNTP and these are known as dye **terminators**. Since fluorescent dyes have different excitation and emission wavelengths, all four reaction mixtures can be loaded into a single lane. A laser scanning system with an optical stage filter wheel and photomultiplier tube moves along the gel. A photomultiplier tube is used to detect the fluorescent emission and converts it into a digital signal. Each time a separate pass-filter is used to detect different colours. The filter passes every 1.5 seconds. Four filters take 6.0 seconds, which is equal to one data point. The data is fed into the computer for sequence analysis. Each fluorescent intensity signal is converted to a corresponding base. This activity is termed base calling. Different colours of the emitted fluorescent dye and the emitted light are detected by a photomultiplier tube. A laser scanning system is attached to the computer to read and store the sequence. This led to the development of semi-automated DNA sequencing machines.

5.5.2 Cycle sequencing

The original chain termination method of DNA sequencing requires cloning of the DNA fragments to be sequenced in an M_{13} vector and subsequent amplification in an *E. coli* host. An alternative to the original method is *cycle sequencing* which involves annealing of the primers with the cloned DNA fragment in a vector, synthesizing the complementary strand in the presence of dNTPs and ddNTPs and DNA polymerase. After one round of extension, the complementary strand is separated from the template by denaturation and the template strand is used for the next round of synthesis. The cycle sequencing has many advantages such as requiring fewer templates for sequencing; more difficult template can be sequenced and working at high temperature reduces the secondary structures. AmpliTAg produced by Applied Biosystem is a mutant version of Taq polymerase in which a single point mutation in the active site (tyrosine is replaced with phenylalanine) helps in less discrimination between dNTP and ddNTP. Due to this, peak height becomes even and base calling becomes more accurate. Moreover AmpliTaq contains a point mutation in the C-terminal which eliminates the 5'–3' exonuclease activity. Since the AmpliTAq is formulated with pyrophosphatase, it converts the ppi formed during extension and subsequently to monophosphate.

5.5.3 Capillary electrophoresis

Beyond certain limits, field strength cannot be increased as it generates heat in the gel. The maximum recorded output using slab gel electrophoresis was

Published by Woodhead Publishing Limited

10,000 bases per day. Due to the introduction of capillary electrophoresis, the speed of sequencing was dramatically increased. In *capillary electrophoresis*, small capillary tubes (50–10 μm, i.d) filled with a sieving matrix are used for electrophoresis. Electrophoresis can be carried out using high field strength, as heat is not generated. Automatic sequencers were developed by ABI for genome project work based on the basis of capillary electrophoresis. Later, multiple capillary tubes were accommodated in a single machine to increase the output of sequencing.

5.6 New sequencing methods

Conventional techniques of DNA sequencing involve high costs, a large quantity of reaction mixtures, manual interference, errors in deducing nucleotide order and long run times. Many advanced automated techniques are being developed to overcome the above-mentioned lacunae. They are characterized by a highly sensitive detection system, bi-directional sequence reads, longer read length, accurate base calling, and low cost and maintenance-free assembly of machines.

5.6.1 Micro-electrophoresis

This is a kind of miniature capillary electrophoresis system. Micro-electrophoretic devices with a cathode and anode were developed. This is a funnel-shaped device with a cathode reservoir of diameter 15 cm at the top and an anode reservoir of 1 cm diameter at the bottom. There are eight microchannels made of fused silica running from the cathode reservoir to the anode reservoir. The surfaces of the microchannels are coated with covalently bound polyacrylamide sieving solutions. The sample is loaded in picolitre quantity at the injector port at the top. When the sample moves down through the microchannels, the laser detector records the fluorescence. The basic principle is directly adopted from the capillary electrophoresis except that miniaturization in volumetric terms is required for sequencing. The microchannel device requires 1,000 times less reagents than capillary electrophoresis. It has been shown that the typical read length is 550 bp with 11.2 cm channel length in 20 min, and the signal to noise ratio is more than 50:1 with uniform signal strength. Separation in the micro-electrophoretic device depends on three parameters such as field strength, fragment size and matrix type. Automation of the microchannels is yet to be demonstrated for commercial application.

5.6.2 Array hybridization sequencing

The target DNA sequence is determined using hybridization between the target DNA and the probe attached to a solid support. The known sequence of short oligonucelotides is attached covalently to a solid support like a glass slide. The Affimetrix Company pioneered fabricating the microarray consisting of 25 bp oligonucelotides on a glass chip in which each oligonucleotide occupies 5 μm^2 with 10,000 copies in each spot. The disadvantage of the sequence by hybridization is that the read length is very small: only 25 bp. Another challenge is cross-hybridization, which poses problems when the repetitive sequences are present. It is also limited by manual sample preparation and PCR amplification of the sample before hybridization to increase sensitivity. This technique is highly successful in genome-wide typing of SNP polymorphism. Comparison of the sequencing technologies is given in Table 5.3.

5.7 Next generation sequencing methods

The next generation sequencing methods are:

1. 454 sequencing technologies
2. ABI-SOLiD sequencing technology
3. Illumina sequencing technology
4. Helicos system
5. Pacific Bioscience technology

Salient features of the next generation sequencing methods

1. Highly streamlined sample preparation.
2. No cloning of the DNA to be sequenced required.
3. PCR-based amplification rather than bacterial amplification.
4. Large read length.
5. Long run time.
6. Reduced manual interference.

5.7.1 454 sequencing technologies

Developed by 454 Life Sciences of Brandford, CT, USA, the 454 Genome Sequencer instrument was marketed by Roche Applied Science and has been

Table 5.3 Comparison of sequencing techniques

Property	Maxam–Gilbert method	Sanger chain termination method	Semi-automated method	Automated capillary sequencing method	Illumina	Solexa	SOLid
Level of automation	Nil	Nil	Fluorescent detection and base calling	Sample preparation, sample loading, signal detection, base calling	Fully automated	Fully automated	Fully automated
High throughput	Nil	Nil	medium	Yes	Yes	Yes	Yes
Accuracy	low	low	high	high	high	high	high
Read length	400 bp–600 bp	400 bp–600 bp	700 bp–800 bp	800 bp–1000 bp	25 bp	35 bp	35 bp

commercially available since 2005. It is one of the first non-Sanger-based sequencing methods to be available on the market. It uses PCR amplification and *pyrosequencing* of the query DNA fragments. The genomic DNA to be sequenced is fragmented and attached to adapter molecules which help the DNA molecules bind to agarose beads. The DNA-attached agarose beads, mixed with Taq polymerase and buffer solution, are added to an oil–water emulsion. Using an oil *emulsion PCR*, the template DNA fragments are amplified to about 10 million copies. In the next step, agarose bead-amplified DNA is placed in picotitre plates which hold one bead per well and these are subjected to pyrosequencing protocol. The PCR amplified beads are placed on glass fibres which are positioned to face the detector system. Polymerase enzyme and primers are added together with any one of the dNTPs. This results in synthesis of complementary DNA strands on the beads. When a particular dNTP is incorporated into the growing strand, it releases a pyrophosphate. The released pyrophosphate is used for the synthesis of ATP and, subsequently, ATP is utilized to generate a light signal from the luciferase enzyme system and the corresponding nucleotide sequence is recorded. The read length is increased to 250 bases. Picolitre titre plates are used to sequence many beads in parallel and it is possible to produce up to one million sequences per day. High cost and lower accuracy of read are the drawbacks of this method. Another limitation of this method is that base calling is not proper when a homopolymer tract is encountered. This does not involve cloning and library construction for *in vivo* amplification of the template DNA fragment. Genomic sequences that cannot be cloned and maintained in bacterial hosts can also be sequenced using this technique

5.7.2 ABI-SOLiD sequencing technology

This was developed in 2007 and works on the principle of genomic library construction and ligation followed by sequencing. It uses DNA ligase for sequencing rather than DNA polymerase. The genome to be sequenced is randomly fragmented and then ligated to the adapter molecules; the adapter-attached molecules are then attached to agarose beads. The bead-captured DNA molecules are amplified using an oil-emulsion PCR. The amplified bead-captured DNA is anchored to a glass slide and flooded with fluorescent-labelled oligonucleotides. If there is complementarity between the template and the oligonucleotide, it is ligated and then two bases are detected at a time. Then the oligonucleotide is cleaved and the next round of ligation commences. Each time two new nucleotides are detected. The read length of this technique is 25 to 35; approximately 40 million beads can be sequenced. The sequencing output of this method is 2 to 4 Giga bases. Since each base is identified twice, the accuracy of this method is high.

5.7.3 Illumina sequencing technology

This was developed by Solexa in 2006. It works on a different principle called reversible terminator nucleotides. A special kind of DNA polymerase is used to incorporate fluorescent-labelled nucleotides. The genomic DNA to be sequenced is fragmented and ligated to adapter molecules on both ends to construct an Illumina-specific adapter library. Then the adapter-ligated DNA fragments are attached to a solid support coated with complementary adapter molecules which help in the anchoring of the DNA fragments. The surface of the support is designed in such a way that it does not absorb fluorescent-labelled nucleotides. Then the PCR amplification of the DNA fragments is performed using the adapter sequence as primer. Approximately 1000 identical copies of the DNA molecules are synthesized and all are laid close together. Since the appearance of each PCR-amplified DNA fragment is like a bacterial colony, these are termed polonies. The size is around one micron in diameter. Then sequencing is carried out by repeated cycles of adding reversible fluorescent-labelled nucleotide and incorporation of the nucleotides to the complementary strand. The fluorescence of the incorporated nucleotides is detected. Then the blockage (terminator group at the 3' end and the fluorescent dye are removed) in the incorporated nucleotide is removed and then the next cycle is started. Likewise, after each cycle of nucleotide incorporation, the fluorescent image is detected. Using this technique, read length is 25 to 35 bases per reaction and it is estimated that approximately 40 million polonies can be sequenced at a time.

Review questions and answers

1. *What is base calling?*

The gel image of the automated sequencer is usually a chromatogram, by using computer software, each peak is converted to a particular base. This step is known as base calling. Usually the automated sequencers are supplied with this software.

2. *How does laser detection increase the accuracy of sequence determination?*

In conventional DNA sequencing methods, radioactivity is used to detect the DNA bands in the autoradiogram. The base sequence has to be deduced manually from the relative banding positions but this resulted in sequence errors. The modern DNA sequencing methods use fluorescent labelling and thus DNA fragments are detected by laser excitation. The emission spectrum is captured and recorded in the computer. A separate software program is used to deduce the base sequence. This greatly reduces the sequencing error.

3. *What are the advantages of Sanger's method of DNA sequencing over the Maxam–Gilbert method?*

Sanger's chain termination method is more amenable to advances because it avoids using toxic chemicals such as piperidine and hydrazine, which are used in chemical degradation method, and also the use of radioactivity for DNA labelling creates health and environmental hazards.

4. *What are the non-Sanger-based DNA sequencing methods?*

- 454 sequencing technologies
- ABI-SOLiD sequencing technology
- Illumina sequencing technology

5. *Capillary electrophoresis has greatly accelerated the speed of DNA sequencing. How?*

Due to the introduction of capillary electrophoresis, the speed of sequencing has been dramatically increased. In capillary electrophoresis, small capillary tubes (50–100 µm, i.d) filled with a sieving matrix are used for electrophoresis. Electrophoresis can be carried out using high field strength, as heat is not generated.

6. *What is the principle of the dideoxy chain termination method?*

ddNTPs are synthetic nucleotides which lack OH group at 3' carbon of the ribose sugar. Whenever ddNTP is added to the growing chain, further elongation will be stopped because incoming dNTP cannot form a phosphodiester bond. Use of dideoxy ribonucleotide is possible only because of the ability of the DNA polymerases to use dNTP and ddNTP as its substrates.

Recommended reading

Ansorge, W.J. (2009) 'Next-generation DNA sequencing techniques', *New Biotechnology*, 25: 195–203.

Bentley, D.R. (2006) 'Whole genome resequencing', *Current Opinion in Genetics and Development*, 16: 545–52.

Ehalich, D. and Matsuderia, P. (1999) 'Microfluidic device for DNA analysis', *TIBTECH*, 315–19.

Hall, N. (2007) 'Advanced sequencing technologies and their wider impact in microbiology', *The Journal of Experimental Biology*, 209: 1518–25.

Hutchison III, C.A. (2007) 'DNA sequencing: bench to bedside and beyond', *Nucleic Acid Research*, 35: 6227–37.

Luckey, J.A., Drossman, H., Kostichka, H.J., Mead, D.A., D'Cunha, J., Norris, T.B. and Smith, L.M. (1990) 'High speed DNA sequencing by capillary electrophoresis', *Nucleic Acid Research*, 18: 4417–21.

Mardis, E.R. (2008) 'Next generation DNA sequencing methods', *Annual Review of Genomics and Human Genetics*, 9: 387–402.

Maxam, A.M. and Gilbert, W. (1977) 'A new method for sequencing DNA', *Proceedings of the National Academy Sciences USA*, 74: 560–4.

Nyren, P. and Lundin, A. (1985) 'Enzymatic method of continuous monitoring of inorganic pyrophosphate synthesis', *Annals of Biochemistry*, 174: 423–36.

Ronagghi, M., Karamohamed, S., Pettersson, B., Uhlen, M., and Nyren, P. (1996) 'Real-time DNA sequencing using detection of pyrophosphate release', *Analytical Biochemistry*, 242: 84–9.

Sanger, F. and Coulson, A.R. (1975) 'A rapid method for determining sequences in DNA by primed synthesis with DNA polymerase', *Journal of Molecular Biology*, 94: 441–8.

Sanger, F. et al. (1977) 'DNA sequencing with chain-terminating inhibitors', *Proceedings of the National Academy of Sciences, USA*, 74: 463–7.

Shendure, J. and Ji, H. (2008) 'Next-generation DNA sequencing', *Nature Biotechnology*, 26: 1135–45.

Shendure, J., Mitra, R.D., Varma, C. and Church, G.M. (2004) 'Advanced sequencing technologies: methods and goals', *Nature Reviews of Genetics*, 5: 335–44.

Web addresses

http://www.solexa.com/.
http://www.454.com/.
http://www.454.com/enabling-technology/the-system.asp.

6

Genome mapping

Science is essentially a cultural activity. It generates pure knowledge about ourselves and about the universe we live in, knowledge that continually reshapes our thinking.

John Sulston

Abstract: Genome mapping is the process of describing a genome in terms of the relative locations of genes and other DNA sequences. Genome mapping is useful for assembling the larger genomes after sequencing. There are two ways to prepare a genome map: the genetic mapping method and the physical mapping method. In this chapter, different genome mapping methods are discussed.

Key words: contig fingerprinting, crossing-over, genetic mapping, molecular markers, optical mapping, physical mapping, radiation hybrid mapping, recombination frequency, restriction mapping, STS content mapping.

Key concepts

- Genes are located on chromosomes. The process of locating the gene order and their relative distances on the genome is known as **genome mapping.**
- There are two kinds of genome mapping techniques: genetic mapping and physical mapping.
- Genome mapping provides guidelines for the reconstruction of the genome sequence after sequencing.
- Genes and molecular markers are used as landmarks to construct a genetic map.
- The recombination frequency (as a percentage) calculated from breeding experiments is used to construct the genetic map.

Published by Woodhead Publishing Limited

- Due to the limited availability of landmarks that could be mapped, a genetic map is a low density map.
- Different types of physical mapping techniques are STS content mapping, radiation hybrid mapping, restriction mapping and FISH mapping.
- Molecular techniques such as Southern hybridization, PCR, restriction analysis, FISH are used to construct a physical map.
- High density genome mapping is possible with physical mapping techniques.

6.1 Introduction

Genome mapping refers to marking specific locations of genes or DNA markers on chromosomes with respect to each other. Genome mapping data helps us to assemble the large DNA fragments that are generated for sequencing. There are two main types of genome mapping techniques: (1) genetic mapping or linkage mapping; and (2) physical mapping. Both methods of genome mapping require a set of markers to map and a methodology to locate the markers on the genome. Different landmarks and techniques are used for genetic mapping and physical mapping techniques.

6.2 Importance of genome mapping in the context of genome sequencing

DNA sequencing method cannot sequence the entire genome in a single reaction. Therefore, the genome is fragmented before sequencing. The ultimate aim of the genome sequencing project is to reconstruct the whole genome after fragmentation and sequencing. After fragmentation of the genome for sequencing, it is a challenge to put all the pieces together in their original order. Some recognizable features are required in order to rearrange the DNA fragments in the same order. Genome maps give a detailed picture of the genome organization in terms of genes, restriction enzymes sites, unique sequences, repeat sequence content, etc. Using these features, the DNA fragments can again be rearranged to fit. Therefore, the **genome map** serves as a guide to arrange the DNA fragments in place and help to reconstruct the original genome after sequencing.

6.3 Genetic mapping

If two genes are located in the same chromosome and are very close, during **crossing-over**, those genes will be inherited together. They are said to be linked. If the two genes are located on the same chromosome but are not close, they get separated during cell division by recombination. Therefore, in subsequent generations either they appear together or separately. These genes are known as partially linked. If the two genes are located on separate chromosomes, they are said to be unlinked. Alleles are different forms of a particular gene which occupy the same loci on the chromosome. During meiosis, homologous chromosomes come together and align themselves at the synopsis stage. The recombination frequency depends on the distance between the two genes. The population of recombinants is analyzed based on the statistical analysis called the LOD score, developed by Newton E. Morton. Links between genes are established when a positive LOD score is obtained, whereas a negative LOD score indicates absence of linkage. An important requirement for mapping is the availability of markers. For genetic mapping, genes with a visible phenotype and DNA markers like RFLP and RAPD can be used. An important requirement for genetic mapping is that the organism must produce many progenies. In plants, *Drosophila*, mouse and yeast, planned crosses are possible, therefore genetic mapping is possible, in them. In humans, planned crosses are not possible and in this case pedigree data are used.

Genes with observable phenotypes are considered good markers for genetic mapping. Usually the selected genes should have multiple alleles. The genes are mapped onto the chromosomes by the breeding technique. Genes with visible phenotypes are limited and it is not possible to map all the genes. Genes that do not show visible phenotypes are difficult to map. These genes are mapped using another type of markers called molecular markers. DNA markers such as RFLP, RAPD, AFLP are used for genetic mapping.

6.3.1 Genetic mapping using genes

Genetic mapping is one of the earliest methods used to map the genes on chromosomes. During meiosis, the non-sister chromatids come together and make **chiasmata** and undergo crossing-over. Crossing-over is a random event and it occurs anywhere along the chromosome. There are some regions on chromosomes where crossing-over occurs more frequently than other regions, which are termed cross-over hot spots. During crossing-over, non-sister chromatids break off and rejoin. This exchange of chromatids results in the exchange of genes.

In genetic mapping, the locations of genes on chromosomes are determined from the recombination frequencies that are calculated from breeding experiments. The number of recombinants to the total number of progenies

Published by Woodhead Publishing Limited

under study will give information about the recombination frequency and usually it is represented as a percentage. The degree of recombination is directly proportional to the distance between genes or their linkage. The recombination percentage depends on the distance between genes on chromosomes. The unit of a genetic map is called a **centiMorgan**, and one centiMorgan corresponds to one percentage recombination frequency. The process by which the linkage status of genes is analyzed is known as linkage analysis. The first genetic map or linkage map was prepared by Thomas Hunt Morgan on *Drosophila*. The linkage map is constructed from a population of recombinants which are obtained after crossing-over. It does not define the physical distance between genes on the chromosome.

Steps in genetic mapping

1. Selection of parents with contrasting phenotypes.
2. Generation of F1 progeny by crossing above selected parents.
3. Generation of F2 progeny by self-pollination.
4. Data collection and calculation of recombination frequency.
5. Construction of genetic map.

6.3.2 Genetic mapping using molecular markers

A molecular marker is defined as any DNA sequence which shows polymorphism and can be detected using a molecular technique. Examples of molecular markers are RFLP, SSR-Microsatellite, Minisatellite, VNTR, AFLP, RAPD, etc. To map these markers on the genome, their inheritance pattern is monitored in subsequent generations. Molecular techniques such as Southern hybridization and PCR are used to detect the polymorphism. The data collected is used to construct the map using computer packages.

6.3.3 Restriction fragment length polymorphism (RFLP)

RFLP is one of the earliest molecular markers developed for genetic mapping. The principle of RFLP markers is that any genomic DNA can be differentiated according to the presence or absence of restriction enzyme sites. Restriction enzymes recognize and cut at the particular site. Due to the accumulation of single nucleotide mutations in genomic DNA, the restriction enzyme sites on DNA change, resulting in a difference in restriction patterns of two closely related genomes. This restriction pattern can be detected using the Southern hybridization technique. Thousands of DNA markers are detected and located throughout the genome. RFLP-based genetic maps have been prepared for many organisms.

Steps in RFLP mapping

1. Selection of parents. They are the first step for DNA markers. They should be selected in such a way that they exhibit maximum polymorphism.
2. Mapping population and size. Different mapping populations can be used for study. Usually it is the F2 population derived from F1 progenies or back-cross population which is selected. The size of the population is also important, usually, more than 1000 is preferred.
3. DNA isolation and restriction digestion. High quality DNA must be isolated from the parents and the F1 and F2 populations before it is used for restriction digestion.
4. Separation of digested DNA. The digested DNA is separated by agarose gel electrophoresis.
5. Southern hybridization. The DNA separated in agarose gel is transferred to nitrocellulose membrane by capillary transfer. The single copy labelled probes such as cDNA or EST are used for hybridization. The autoradiogram developed reveals the banding position. Scoring for the presence and absence of DNA bands is done for the parents, the F1 progenies and F2 progenies.
6. Construction of RFLP map using computer program. Based on the data collected, the likelihood of linkage and the absence of linkage between markers can be calculated to obtain the LOD score in order to construct the genetic map. Since the data is large, a computer program has been developed to do it. The most popular program for genetic map construction is Mapmaker.

6.4 Genetic mapping in humans

The above-described genetic mapping technique cannot be applied directly to humans. The essential requirements for *genetic mapping* include: (1) availability of genes or DNA markers; (2) planned crossing between homozygous dominant and recessive parents; and (3) a large number of progenies. Owing to societal limitations, all the above said requirements cannot be met in humans. Therefore, genetic mapping in humans is challenging. Progress that has been made in human genetic mapping has helped assemble human genome sequence data. The method adopted for genetic mapping in human beings is known as a **pedigree analysis**. This is the process by which information is collected from family records. Earlier, human gene mapping was done with a view to locate the genes on the chromosome so that they could be cloned for further analysis of their mode of transmission through the generations. The genes that were mapped were the ones implicated in genetic diseases which are either X- linked or autosomal. In the genomic era the genetic maps constructed using these disease genes were used to arrange the clone contigs. For genetic mapping using pedigree analysis, the data should be collected from all four grandparents, parents and a minimum of six siblings. The Human Polymorphism Study Centre, the Centre d'Etude du Polymorphisme Humain (CEPH) was founded by a French immunologist in 1984, using the Nobel Prize money. Transformed cell lines of reference pedigrees from 61 human families from throughout the world are maintained in this centre. This centre was established with the aim of mapping and cloning human disease genes and finding a cure for the diseases that were due to faulty genes.

6.5 Physical mapping methods

Physical mapping involves locating the DNA sequences directly on the chromosome or large genomic clones. The unit of measurement is the base pair. **Cytogenetic mapping** is one of the earliest developed physical mapping methods, but it was not sufficiently accurate to give a high resolution map of the chromosomes. Physical mapping methods were explored to prepare high density maps. Physical mapping methods that were developed were able to give a very high density map (1 marker/100 kbp) that is needed for genome projects to prepare finished genome sequences. The markers for physical mapping include ESTs, genome-wide unique DNA sequences, and STS markers. The **STS (Sequence Tagged Site)** marker is any unique DNA sequence whose sequence is known and can be mapped using a particular molecular technique.

6.5.1 Cytogenetic mapping

Cytogenetics refers to the study of structure and functions of chromosomes in a cell. Initially, the interest in cytogenetics arose when the fact that the chromosomes were the carriers of genetic material came to light. During cell division, the chromosomes undergo vivid structural changes. The chromosomes can be arranged according to their size from the largest to the smallest. This is known as **karyotyping**. Because of their organization and the nature of DNA sequences, chromosomes are known to bind differentially to fluorescent organic dyes. When the chromosomes are mixed with organic dyes, they exhibit a characteristic banding pattern. The most commonly observed banding pattern is G-banding. G-banding is done with the help of Giemsa. Subsequently, many other banding techniques were developed including Q-banding with Quinacrine. The banding pattern of the chromosomes gives a macroscopic view of the organization of chromosome, and they have proved useful in the clinical diagnosis of many genetic diseases and chromosomal abnormalities. *In situ* hybridization was reported by three groups of scientists in 1969 (Gall and Pardue, Buongiorno-Nardelli and Amaldi and John et al.). They used labelled RNA as a probe to identify the DNA on a chromosome. The principle of *in situ* hybridization relies on nucleic acid annealing between the probe and the DNA strand present in the chromosome. Initially radio-labelled probes were used and the means used to view the results was autoradiography. Owing to its inherent health hazards and poor resolution, the probe labelling technique was changed to fluorescent labelling. The method is popularly known as FISH.

6.5.2 FISH in genome mapping

Fluorescent *in situ* Hybridization (FISH) is a technique to locate a gene or DNA sequence directly on the chromosomes. It is a physical mapping technique. The metaphase chromosomes are used for *in situ* hybridization. Due to their highly compact nature, very closely located genes or DNA sequences are mapped on the same location of the metaphase chromosomes. This results in poor resolution of the cytogenetic map. Further improvements in cytogenetic mapping techniques have been made using less condensed chromosomes, such as pachytene chromosomes or very large DNA sequences, which are combed on a glass slide and are known as fibre FISH.

Steps in FISH

1. *Fixation and denaturation.* An individual chromosome or whole chromosomes at the metaphase cell division stage are separated

---

from the cells and put onto the microscope slide. They are kept on the slide by adding fixatives. The proteins are then denatured by adding formaldehyde.

2. *Probe preparation.* Single-stranded cDNA or EST is used for labelling and fluorescent dye is covalently attached to the DNA.

3. *Hybridization.* The chromosome or the DNA attached to the glass slide is denatured with formaldehyde to render it single stranded. The labelled probes are flooded on the chromosome and incubated. The unhybridized labelled probes are washed.

4. *Visualization.* The slide from step 3 is visualized under a fluorescent microscope (Figure 6.1).

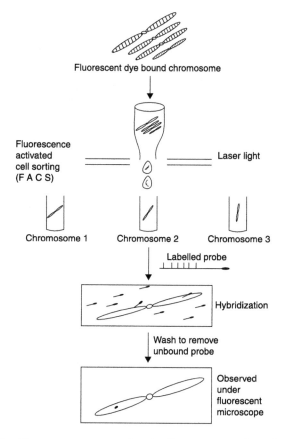

Figure 6.1   Chromosome separation by fluorescent activated cell sorting (FACS), followed by fluorescent probe hybridization to locate the position of the DNA on the chromosome.

Applications of FISH

1. Medical diagnosis of genetic diseases related to chromosomal abnormalities.
2. Ordering genes or DNA sequences directly on the chromosome, i.e. gene mapping.

Advantage of FISH

1. It provides a direct relationship between the probe sequence and chromosomes.

Limitations of FISH

1. Skilled personnel are required to handle the chromosomes.
2. As far as genome mapping is concerned, FISH provides only a low resolution map. Although advances have been made to increase the resolution of the FISH map, the maximum possible resolution is one marker/Mbp.

6.5.3 Restriction mapping

Restriction mapping is a physical mapping technique which is used to determine the relative location of restriction sites on a DNA fragment to give a restriction map. Restriction enzymes are endonucleases that recognize specific sequences on DNA and make specific cuts. These enzymes are produced by bacteria to protect themselves from bacteriophage attack. These are classified into four groups: type I, type II, type III and type IV. Type II enzymes are most frequently used for molecular techniques. Restriction mapping involves the positioning of relative locations of restriction sites on a DNA fragment. Some enzymes recognize four bases as restriction sites and make frequent cuts and generate relatively smaller fragments, while some enzymes recognize six bases or eight bases and make rare cuts to generate larger fragments (e.g. the NotI recognition sequence is GGCGCGCC). When a large genomic DNA is digested with a frequent cutting enzyme, millions of DNA fragments are generated. Separation of these fragments by conventional agarose gel electrophoresis results in no discrete bands. Therefore, rare cutting enzymes are useful for genome mapping as they generate relatively less DNA fragments. After digestion, DNA fragments are separated in agarose gel using a special technique called Pulse Field Gel Electrophoresis (PFGE). Using this technique, DNA fragments of up to 10 Mbp can be separated as against less than 40 kbp which can be separated using conventional agarose gel electrophoresis (Figure 6.2).

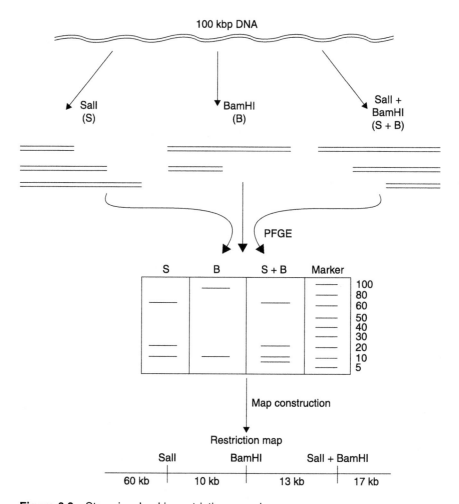

Figure 6.2   Steps involved in restriction mapping.

Published by Woodhead Publishing Limited

Steps in restriction mapping

1. Preparation of DNA for restriction analysis.
2. Restriction digestion of DNA.
3. Separation of restricted DNA.
4. Collecting data.
5. Construction of restriction map.

Restriction mapping of the genome is useful only for smaller genomes such as viruses and bacteria. Restriction digestion of the large genomes produces many DNA fragments. When these fragments are separated in agarose gel, DNA fragments cannot be viewed as discrete bands. Another version of restriction mapping called **optical mapping** is used for larger genomes. In this method, DNA fragments are not separated in agarose after digestion. Restriction digestion is done by placing the isolated chromosome in agarose solution on a microscopic slide and allowing a gel to form. When the agarose forms the gel, the DNA is stretched. The restriction enzyme is flooded on the agarose gel containing DNA and the mixture is incubated. Then the restricted DNA is observed under a high power microscope and the relative location of the restriction sites are visualized as gaps. From the location of gaps, the restrictions sites are mapped.

6.5.4 STS content mapping

STS content mapping is one of the best physical mapping methods by which high resolution genome mapping is achieved. An important requirement for physical mapping is that the genomic DNA is fragmented so as to produce DNA fragments in an overlapping manner. For STS content mapping we require to have some recognizable unique sequences. A commonly used STS marker is the **Expressed Sequence Tag (EST)**, which is unique and is produced by partial sequencing of cDNA library clones. Due to the development of high capacity cloning vectors and PCR-based identification methods, STS content mapping became the most effective method for physical mapping.

Steps in STS content mapping

1. Fragmentation of genome: Fragmentation of the genome into overlapping clones is usually done by using rare cutting enzymes so as to produce large DNA fragments.
2. Preparation of a large fragment library: The fragments thus generated are ligated to cloning vectors of high capacity like YAC.
3. Overlapping clone identification: Two strategies are used to identify overlapping clones: one is **clone fingerprinting** and the other is chromosome walking. In the clone fingerprinting method, overlapping clones are identified based on common patterns of restriction enzyme fragments or repeat content observed between clones. In chromosome walking, the overlapping clones are identified based on hybridization of two overlapping clones to a particular STS marker.
4. Preparation of STS content map: Arranging the overlapping fragments based on the presence or absence of a signal for a particular probe.

6.5.5 Radiation hybrid mapping

This is a type of physical mapping technique. It generates a high resolution map of the genome. Rodent cells have the ability to maintain chromosomes derived from other species as part of their chromosomes. It is possible to induce chromosomal breakage by exposing the cells to X-rays. The breakage depends on the intensity and the duration of the irradiation. The irradiated cells with broken chromosomes are fused with rodent cells and grown on a HAT medium which allows only hybrid cells to grow. This gives a panel of radiation hybrids. The landmarks that can be mapped on the radiation hybrids include STS, EST, etc. Even non-polymorphic markers can be mapped using radiation hybrid mapping technique (Figure 6.3). The presence or absence of markers on a particular *radiation hybrid* is identified by PCR amplification. The linkage between two markers is determined by calculating the breakage percentage based on PCR amplification. When two markers are located close together, they will be retained together during irradiation, otherwise they will get separated. Breakage depends on the distance between markers. A particular radiation hybrid will show PCR amplification for two markers if they are located close to each other. The

Published by Woodhead Publishing Limited

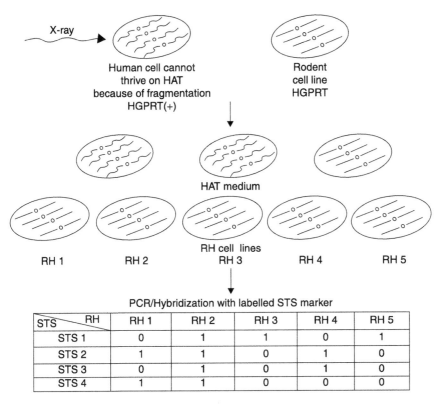

Figure 6.3 Radiation hybrid mapping. Chromosomes are fragmented using radiation energy; the fragmented chromosomes are ligated to the rodent chromosomes to propagate in rodent cell lines as radiation hybrids; each RH cell line is examined for the presence of an STS marker. The relative location of the STS markers and their overlapping clones are identified using molecular techniques like PCR.

LOD score is calculated, the order and distance between the markers are mapped. This technique is useful when STS content mapping using YAC clones is not feasible, owing to poor maintenance of certain parts of the chromosome such as GC-rich regions and terminal parts of the chromosomes, etc.

Steps in RH mapping

1. Irradiation of human fibroblast cells.
2. Fusion of irradiated cells with rodent cells.

> 3. Selection of fused cells on HAT medium.
> 4. PCR amplification with specific primers using hybrid chromosomes.
> 5. Construction of RH map.

When the Human Genome Sequencing Project started in 1990, scientists opted for map-based genome sequencing strategy. Therefore, one of the objectives of the Human Genome Project was to develop high density mapping methods. Targets were set for genetic mapping and physical mapping. Many new mapping methods such as optical mapping, fibre FISH and radiation hybrid mapping were developed. A comparison of the different mapping methods is given in Table 6.1.

Review questions and answers

1. *Define a genome map.*

A genome map is the graphical description of the location of genes and DNA markers on the genome of an organism.

2. *What are the different types of genome mapping methods?*

- Genetic mapping
  - Mapping using genes
  - Mapping using molecular markers
- Physical mapping
  - Restriction mapping
  - Cytogenetic mapping
  - STS content mapping
  - Radiation hybrid mapping

3. *How does a genetic map differ from a physical map?*

A genetic map is constructed using recombination frequency calculated from the progenies and it is an indirect method of locating the positions of genes or DNA markers. The unit of measurement is cM, whereas physical mapping pertains to locating the position of DNA sequences directly on the chromosome of a large DNA fragment. The unit of measurement is the base pair.

Table 6.1 Comparison of different mapping methods

Sl. No.	Genetic mapping			Physical mapping		
	Gene mapping	Molecular markers	Cytogenetic mapping	Restriction mapping	STS content mapping	Radiation hybrid mapping
Mapping method	Gene mapping	Molecular markers	Cytogenetic mapping	Restriction mapping	STS content mapping	Radiation hybrid mapping
Experimental protocol	Breeding Pedigree analysis Conjugation	Southern hybridization	FISH PCR	Restriction analysis	PCR	PCR/Nucleic acid hybridization
Landmarks to be mapped	Genes	RFLP, RAPD, SSR	Gene or unique DNA sequence	Restriction sites	Any unique DNA sequence	Any unique DNA sequence
Resolution	low	medium	medium	medium	high	medium
software	–	Mapmaker	–	FPC (Finger printing contig)	IMP (Integrated Mapping Package)	RHMAP version 2.01
Application to genome sequencing	low	medium	medium	medium	high	high

Published by Woodhead Publishing Limited

4. *What are the markers for genetic mapping and physical mapping?*

Markers for genetic mapping
- Genes with visible phenotype
- Molecular markers such as RFLP, AFLP and SSR

Markers for physical mapping
- Restriction enzyme sites
- Expressed sequence tags
- cDNA clones
- Any unique DNA sequence

5. *Clone contig mapping is more useful than other physical mapping methods for genome sequencing. Justify the statement.*

Clone contigs are prepared from BAC clone libraries. The BAC clones can also be used both for physical mapping and sequencing.

6. *High density mapping is helpful in genome sequence finishing. Validate the statement.*

Eukaryotic genomes are large and they contain a sizable portion of repetitive sequences. Genomic DNA is fragmented into a smaller DNA before sequencing; clone assembly produces contigs but not the entire reconstructed genome, i.e. it has many gaps. To close the gaps, it is necessary to have high density maps.

7. *STS map is one of the highest density physical maps. How is this possible?*

Thousands of STS markers are available to map on the BAC library clones using highly sensitive clone fingerprinting techniques. Therefore, it is possible to construct a map with one marker in every 100 kbp of the genome.

8. *Restriction mapping is not a better physical mapping technique for larger genome mapping. Why?*

When the size of the genome is too large, the number of fragments produced by restriction digestion will increase, and the restricted fragments separated after gel electrophoresis cannot be viewed as discrete bands.

Recommended reading

Agarwala, R., Applegate, D.L., Maglott, D., Schuler, G.D. and Schaffer, A.A. (2000) 'A fast and reliable radiation hybrid map construction and integration strategy', *Genome Research*, 10: 350–64.

Dib, C. et al. (1996) 'A comprehensive genetic map of the human genome based on 5,264 microsatellites', *Nature*, 380: 152–4.

Hudson, T.J. et al. (1995) 'An STS-based map of the human genome', *Science*, 22: 1945–54.

International Human Genome Mapping Consortium (2001) 'A physical map of the human genome', *Nature*, 409: 934–41.

Lander, E.S. and Botstein, D. (1986) 'Mapping complex genetic traits in humans: new methods using a complex RFLP linkage map', *Quantitative Biology*, 51: 49–62.

Meksem, K. and Kahl, G. (2005) *The Handbook of Plant Genome Mapping: Genetic and Physical Mapping*, Berlin: Wiley-VCH Verlag GmbH and Co KGaA.

Nelson, W., Sodulund, C. and Mott, R. (2009) 'Integrating sequence with FPC fingerprint maps', *Nucleic Acid Research*, 37: 30.

Schuler, G.D. et al. (1996) 'A gene map of the human genome', *Science*, 274: 540–6.

Zhang, P., Ye, X., Liao, L., Russo, J.J. and Fisher, S.G. (1999) 'Integrated mapping package: a physical mapping software tool kit', *Genomics*, 55: 878–87.

Web addresses

http://www.ncbi.nlm.nih.gov/genemap99.
http://www.ncbi.nlm.nih.gov/mapview/.
http://www.ornl.gov/sci/techresources/Human_Genome/research/mapping.shtml.

7

Genome sequencing methods

Abstract: Genome sequencing involves revealing the order of bases present in the entire genome of an organism. Genome sequencing is backed by automated DNA sequencing methods and computer software to assemble the enormous sequence data. It can be divided into four stages: (1) preparation of clones comprising the entire genome of an organism; (2) collection of DNA sequences of clones; (3) generation of contig assembly; and (4) database development. In this chapter, two popular genome sequencing methods – whole genome shotgun sequencing and the clone-by-clone method – are discussed in detail.

Key words: clone-by-clone method or hierarchical method, genome sequencing, high throughput sequencing, sequencing clones, source clones, whole genome shotgun sequencing.

Key concepts

- Genome sequencing refers to sequencing the entire genome of an organism.
- Many high throughput sequencing and data handling technologies have been developed.
- Major genome sequencing methods are the clone-by-clone method and the whole genome shotgun sequencing.
- The clone-by-clone method of sequencing works well for larger genomes like eukaryotic genomes but it requires a high density genome map.
- Whole genome shotgun (WGS) sequencing does not require a genome map. The WGS method is a faster method of sequencing but is not suitable for larger genomes like eukaryotic genomes as they have a number of repetitive DNA sequences in which the assembling process is challenging.
- Hence, to speed up the genome sequencing process, advantages of both methods are used.

Published by Woodhead Publishing Limited

7.1 Introduction

The genome sequencing of many organisms resulted in the development of a number of branches such as genomics, proteomics, and metabolomics, etc. **Genome sequencing** refers to sequencing the entire genome of an organism, instead of sequencing it gene by gene. Sequencing the entire genome will provide a wealth of data which can be studied completely. Studying the functions of genes one by one is time-consuming, it leads to redundant work by many scientists, slow progress, and only incomplete information is obtained. Due to the advances in DNA sequencing technology, genome sequencing has been made possible. Genome sequencing is easy with prokaryotic organisms because their genome is small and contains very few or no repetitive sequences. But sequencing the eukaryotic genome is difficult due to its large size and the large number of repetitive sequences. However, improvements in sequencing technology and computerized data handling have paved the way for sequencing of even very large genomes like the human genome. After the Human Genome Project was initiated, many high throughput automated sequencing techniques were developed and executed. **Nucleotide** sequence is the highest resolution map of a genome. It provides comprehensive information about genes and their related regulatory elements and other features of a genome.

Reasons for proceeding to genome sequencing

1. Genes do not work alone in cells; genome sequencing helps us to understand the functions of all the genes which are present in the cells.
2. Gene function is not directly controlled by the **promoter** alone; it is controlled by many other regulatory elements such as the response elements, the enhancers, the silencers, etc. Whole genome sequencing gives information about the related DNA elements involved in gene expression.
3. The presence of vast amounts of non-coding DNA in the eukaryotic genome without any known obvious function is still an enigma. Genome sequencing and its characterization will provide some clue to the function of non-coding DNA.
4. The genomes of individuals of a single species are similar but not identical, in the case of humans, about 2 per cent of the genomic DNA sequence varies among individuals.

Published by Woodhead Publishing Limited

Genome sequencing was started with the ultimate aim of sequencing the human genome. But many model organisms' genomes were sequenced to test the feasibility and efficiency of techniques like mapping, sequencing, and assembling, etc. There are two methods of genome sequencing. The first method is the hierarchical method or the clone-by-clone method or the map-based or BAC-based method. The second method is whole genome shotgun sequencing. Whole genome shotgun sequencing works well with smaller prokaryotic genome sequencing where repetitive sequences are few, hence, assembly after sequencing is simple.

7.2 The clone-by-clone genome sequencing method

This method is also known as the top-down method of genome sequencing. This method involves preparing a high density map using genetic and physical markers. It works on the principle of the top-down approach, i.e. first mapping the genome and then sequencing it (Figure 7.1). This was the method adopted by the publicly funded HGP consortium.

Steps in the clone-by-clone method of genome sequencing

1. Preparation of BAC clone library.
2. Preparation of clone fingerprint.
3. BAC clone sequencing.
4. Sequence assembly.

7.2.1 Preparation of the BAC clone library

The high quality genomic DNA is partially digested or randomly fragmented to produce overlapping fragments. The genomic DNA is fragmented into large DNA fragments of 150 kbp size and cloned into BAC vectors. These are transformed into E. coli host and maintained as a *BAC library* for further use. This library is termed the source clone library. It minimizes the number of clones to be maintained for each genome.

7.2.2 Preparation of the clone fingerprint

These BAC clones are rearranged according to the overlapping nature of the DNA markers such as the STS marker or restriction sites. The length of

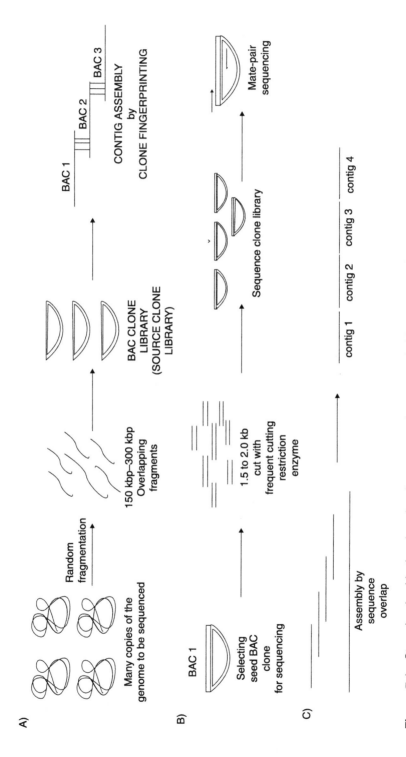

Figure 7.1 Steps involved in the clone-by-clone sequencing approach. Many copies of the genome to be sequenced are randomly fragmented to obtain larger overlapping clones and used for sequencing.

the BAC clones that constitute the entire genome is known as the tiling path. Each recombinant BAC DNA fragment is separated and digested with a frequent cutting enzyme and separated on agarose gel. The pattern of the agarose gel fragment is compared across the entire BAC fragment. Since the original large DNA fragments are overlapping, the sequences overlap to a certain degree. This results in a restriction pattern and also overlapping. This restriction pattern will be the fingerprint for each DNA sequence since the overlapping clones can be hooked together based on their overlapping restriction enzyme fingerprint. The individual BAC clones are digested further to yield small fragments of 1–1.5 kbp and cloned into **plasmid** vectors. These plasmid-based libraries of individual BAC clones are sequenced from both ends. Each clone is sampled many times and sequenced. This ensures that the entire DNA present in the BAC is sequenced. The approach prepares an outline of the genome. When the genome sequence is completed based on the overlapping sequence of small inserts, they are put together.

7.2.3 BAC clone sequencing

After anchoring the BAC clones according to their physical mapping details, they are selected for complete sequencing. The selection of BAC clones for the sequencing process is known as seed BAC clone selection. The BAC clones are stored in glycerol stocks. Before proceeding to sequencing, it is important to verify the BAC clones. This is done by locating one or two markers by PCR amplification or hybridization with labelled probes performing FISH to confirm its physical location on the BAC clone sequence.

7.2.4 Sequence assembly

Each sequence is obtained from sequencing four independent clones and finally they are entered as a consensus sequence to obtain a finished sequence. Therefore, the fold coverage is approximately 8–10 times. The clone-by-clone or BAC-based sequencing method was undertaken by the publicly funded Human Genome Project. They started by preparing a high density map using all the possible mapping techniques. This was done because this reduces the error rate of sequencing and an accurate sequence assembly is possible after sequencing, even with a large number of repetitive sequences. Otherwise it is better to map first and sequence later.

Published by Woodhead Publishing Limited

Advantages of the clone-by-clone method of genome sequencing

1. Since high density maps are available, making the genome assembly is easy.
2. The finished genome sequence will be completed in an accurate way by using maps.

Disadvantage of the clone-by-clone method of genome sequencing

It requires a high density map. Preparing a high density map is a time-consuming and costly process.

7.3 The whole genome shotgun sequencing method

The whole genome shotgun (WGS) method was proposed as an alternative method to the map-based clone-by-clone strategy for human genome sequencing, which was adopted by the publicly funded International Human Genome Sequencing Consortium (IHGSC). The WGS method involves less cost and less time, compared to the clone-by-clone method. The Lambda phage genome (~49 kbp) was the first completely sequenced genome using the WGS method. Other viral genomes like the smallpox virus were sequenced using this method. In 1995, the WGS method was used to sequence the 1.8 Mbp size *Haemophilus influenzae*, which was the first bacterial genome that was sequenced using this method. Using the TIGR EST assembly algorithm, a random sequence of this bacterium was successfully assembled. Subsequently, this method was successfully applied to many other genomes of small to medium size. Many viral and bacterial genomes have in fact been completely sequenced by the WGS method. Since smaller genomes do not have repeat sequences, fewer problems were faced during sequence assembly phases, but when it comes to large eukaryotic genomes with repetitive sequences, this method may not work. *D. melanogaster* was selected as a model eukaryotic genome to test the suitability of the WGS strategy. *Drosophila melanogaster* genome sequencing was successfully done using WGS method. When the Human Genome Sequencing Project started, the adaptability of this method to sequence the huge human genome was doubted by many scientists. Many scientists suggested adopting the map-based approach to sequence the human genome.

Published by Woodhead Publishing Limited

Steps in the whole genome shotgun sequencing

1. Isolation of high quality genomic DNA.
2. Random fragmentation of genomic DNA by ultrasonication.
3. Size fractionation using agarose gel electrophoresis.
4. Library construction.
5. Paired-end sequencing.
6. Assembly.

The first step in whole genome shotgun sequencing is the generation of small overlapping random fragments which cover the genome of an organism. The highly purified genomic DNA is randomly fragmented into smaller fragments and then selected on size after agarose gel electrophoresis. Two different sizes of DNA fragments – long insert (2–25 kbp) and short insert (0.5–1.2 kbp) – are selected from the agarose gel. The long inserts are cloned in phage or cosmid vectors and the short inserts are cloned in plasmid vectors. The short insert clone library is used for sequencing. Clones are selected from the short insert clone library and sequenced from both the ends. Since the size of the short insert is kept small (< 1.2 kbp), sequencing of this insert from both ends gives read lengths which overlap. This is known as a paired-end sequence or mate pair sequencing. Large numbers of clones from the short insert library are sequenced which results in huge nucleotide sequence data. Each of the genomes will be covered more than 10 times. Use of both short insert and long insert clones increases the efficiency of sequence assembly (Figure 7.2).

Advantages of whole genome shotgun sequencing

1. It does not require any genome map.
2. Less time consuming.
3. Money is saved.

Disadvantages of whole genome shotgun sequencing

1. When this method is applied to eukaryotic genomes, assembly is difficult as they have quite a lot of repetitive DNA sequences.
2. The finished genome sequence using this method is not accurate.

Published by Woodhead Publishing Limited

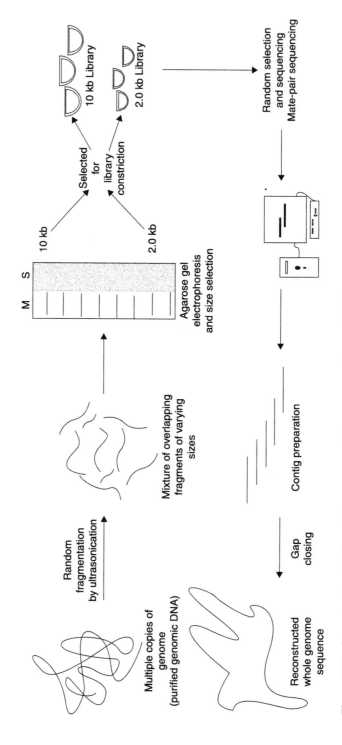

Figure 7.2 Steps involved in the whole genome shotgun sequencing methodology.

Similarities between the clone-by-clone method and the WGS sequencing method

1. Genome is fragmented into overlapping DNA fragments.
2. Cloned into cloning vectors and amplified in *E. coli* host.
3. Samples from different ethnic peoples were collected and randomly selected for sequencing.
4. Assembling algorithms were used to reconstruct the original genome.

Table 7.1 Differences between genome sequencing methods

Sl. No.	Clone-by-clone method	WGS sequencing method
1	Man-based method; requires high density map of the genome	Does not require genome map
2	Genome is fragmented into ~150 kbp DNA fragments which are cloned into high capacity vectors like BAC	Genome is fragmented into small overlapping fragments of size 2–10 kbp and cloned into plasmid vectors
3	Each BAC clone is fragmented further into 2 kbp small fragments and sequenced	Each is sequenced from both ends to produce a mate-pair sequence
4	Clones are anchored using mapping features already available	Clones are assembled based on the overlapping sequence of the mate-pairs
5	Number of gaps produced is less	More gaps are produced
6	Large genomes with repetitive sequences can be correctly assembled using map data	It is better suited to smaller genomes without repetitive sequences
7	Costly and time-consuming as detailed genome mapping requires resources and skilled personnel	Less costly and rapid method as it does not require detailed genome mapping resources and skilled personnel
8	Assembling is easy and works well with simpler computer software	Computationally intensive
9	Sequencing and assembling can be performed in many centres as each centre can select one BAC and work on it.	Sequencing can be done in many centres but assembling should be performed in a particular centre

7.4 Error control in genome sequencing

The originally developed DNA sequencing protocols relied on manual sequence interpretation. Therefore, the error rate was very high. The current genome sequencing techniques do not involve any manual interference for sequence determination as they use capillary electrophoresis and dideoxynucleotides labelled with fluorescent dyes. Error control is a very important part of the genome sequencing. Unless the errors are minimized to some acceptable level, the generated reference sequence will not be of any use. The raw sequence data obtained from the automated advanced sequencing machines also show sequencing errors. The output from the sequencer is a chromatogram showing the intensity of the peaks against time.

Errors can be due to the experimental handling of the reaction mixtures, electrophoretic separation, compression artifacts, signal dropouts, etc. This results in wrong base calling. The correct nucleotides that can be deduced from a chromatogram vary from 600–1300 bp per sequencing reaction. The initial trace will have a high error rate as it varies from 3–8 per cent up to 50–70 bases, followed by high quality data with an error rate of < 1 per cent. Again the signal quality fades away at the end; the error rate rises to 10–20 per cent (Figure 7.3). The kinds of errors introduced in the sequences are of three types: insertions, deletions and mismatches. To avoid the wrong base call, computer algorithms have been developed to take many other characters (such as peak height, peak shape) apart from intensity of the peak. For each base call, a certain probability value is attached, which is denoted by the 'p' value. The 'p' value ranges from 0–1 ($0 \leq p \leq 1$). A 'p' value of 1 means that the called base is wrong. The computer program used for this is Phred, which assigns a probability value to each base call. A Phred score is allotted to each base called. It is calculated from the peak shape, and height in the electrophorogram of the automatic sequencer.

The DNA fragments used for sequencing are cloned in a cloning vector and amplified in a biological host. Initial and end sequences of each clone will be of the vector sequence, that have to be removed. Error is also introduced when the cloned sequence replicates in the host, this results in a small error called point mutations single nucleotide polymorphism (SNP). These SNPs do not cause any major problem. Another major error in the sequence determination is due to recombination of the cloned DNA with the host DNA. The sequence detected with this kind of sequence is called a chimera. This type of error creates a major sequence error rate. The acceptable error rate for the genome sequence is 99.99 per cent (only one

Published by Woodhead Publishing Limited

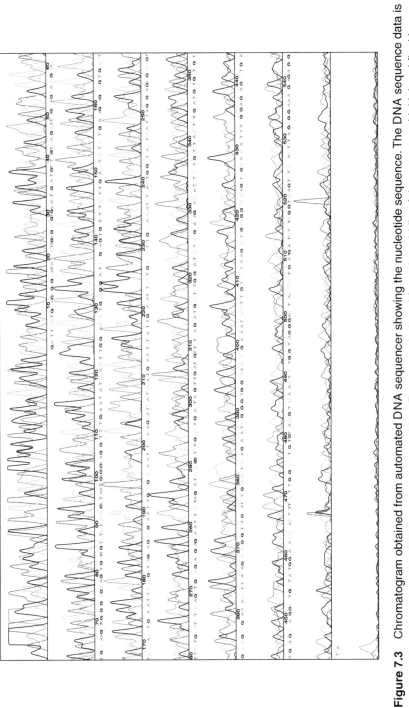

Figure 7.3 Chromatogram obtained from automated DNA sequencer showing the nucleotide sequence. The DNA sequence data is obtained from the automated DNA sequencing machine. Bases are deduced from the middle of the output and initial and final bases are not reported due to poor signal quality which results in a high error rate.

error in a 10,000 bp sequence). This is the world standard for genome sequence fidelity which was agreed in a meeting in Bermuda by the genome sequence principal investigators in 1997; therefore it is called the Bermuda standard.

Review questions and answers

1. *What are the commonly used methods for genome sequencing?*

- Clone-by-clone method or hierarchical method.
- Whole genome shotgun method.

2. *What is a source clone library?*

Large insert clones prepared in BAC vector which are used for clone assembly and source for sequencing in the clone-by-clone method.

3. *What is a tiling path?*

The length of BAC clones used to cover the entire genome of organism.

4. *Why is not possible to use only the WGS method for eukaryotic genome sequencing?*

Eukaryotic genomes have repetitive DNA sequences; if WGS is followed, genome assembly will be difficult.

Recommended reading

Olson, M.V. (2001) 'Clone by clone by clone', *Nature*, 409: 816–18.
Roach, J.C., Boysen, C., Wang, K. and Hood, L. (1995) 'Pair wise end sequencing: a unified approach to genomic mapping and sequencing', *Genomics*, 26: 345–53.
Staden, R. (1979) 'A strategy of DNA sequencing employing computer programs', *Nucleic Acid Research*, 6: 2601–10.
Sterky, F. and Lundeberg, J. (2000) 'Sequence analysis of genes and genomes', *Journal of Biotechnology*, 76: 1–31.
Venter, J.C., Adams, M., et al. (1998) 'Shotgun sequencing of the human genome', *Science*, 280: 1540–2.
Venter, J.C., et al. (2001) 'The sequencing of the human genome', *Science*, 291: 1301–51.

Published by Woodhead Publishing Limited

Venter, J.C., Smith, H.O. and Hood, L. (1996) 'A new strategy for genome sequencing', *Nature*, 381: 364–6.

Weber, J.L. and Myers, W.E. (1993) 'Human whole-genome shotgun sequencing', *Genome Research*, 7: 401–9.

Web address

http://www.gene.ucl.ac.uk/hugo/bermuda2.html.

Published by Woodhead Publishing Limited

8

Genome sequence assembly and annotation

Abstract: The sequencing technology supports sequencing of small DNA fragments only. Therefore, sequencing of the genome is carried out after fragmenting the genomic DNA into small pieces. Reconstructing the original genome sequence from millions of small pieces is a daunting task. In this chapter, the various steps involved in the genome sequence assembly are discussed. It is not possible to reassemble the original genome sequence; instead, the assembled genome sequence shows gaps which are later closed using different gap closing methods. The assembled genome sequence has to be annotated to use it for further study.

Key words: annotation, Bactig, contig, depth of coverage, draft sequence, finished sequence, supercontig.

Key concepts

- Genomes are fragmented into overlapping DNA fragments before sequencing.
- After genome sequencing, the DNA fragments are arranged to reconstruct the original genome sequence.
- The reconstructed genome sequence show gaps and these gaps are closed by gap closing methods.
- The process of assembling the original sequence from the overlapping clones is called genome assembly.
- Computer software programs are used to assemble the overlapping clones.

8.1 Introduction

A genome is an ordered arrangement of huge nucleotide sequences. For genome sequencing, the genome is randomly broken down into small

Published by Woodhead Publishing Limited

overlapping fragments. Once the genome is broken down, it is very difficult to reorder it again unless some landmarks are used. No sequencing method has yet been discovered which could determine the sequence of the entire genome at a stretch. Instead, each sequencing reaction produces approximately 600–700 **nucleotides**. The length of sequence that can be determined from a single reaction is known as the read length. Reconstructing the original genome from these small read lengths is a big challenge. The overlapping clones can be arranged using various clone assembly techniques. This process is known as **genome assembly**, and it is more like a jigsaw puzzle.

The clones have to be ordered and oriented using overlapping nucleotide sequences. This is not performed manually as it requires that millions of reads have to be matched. Computers play a major role in this assembly process and algorithms have been developed to achieve this. The genome sequence assembly is affected by the presence of errors in the input files and the occurrence of repetitive sequences in the genome. The genome sequence assembly algorithms produce assembled genome sequences with gaps. Initially, assembly of the sequence gives a rough draft of the genome. The draft genome sequence has base ambiguities, gaps and the coverage is low. The draft genome sequence requires further analysis and the sequence gap and physical gaps must be closed using targeted library sequencing strategies. The final sequence generated is known as the finished genome sequence which has few gaps and the sequence represented is the consensus sequence with fewer ambiguities.

8.2 Clone contig assembly

Contig assembly is an important step in genome assembly. The word contig was first described in a paper in 1980 by R. Staden (see Recommended reading). The set of the overlapping DNA sequence of DNA fragments is known as a contig. Contig mapping is a process by which overlapping clones are assembled to sequence that overlap. This involves arranging the contigs in order and orientation. Clone contigs can be automatically assembled using their BAC-end sequences.

The basic principle of sequence assembly is that the overlapping clones share the same nucleotide sequence as the genomic DNA. Usually many copies of the genome are fragmented which cover the entire genome 8 to 10 times. Each fragment is cloned in a vector and sequenced from both ends to produce a sequence length of approximately 600–700 bp. The sequence from both ends of DNA fragment is called a mate-pair. The sequence from both ends of insert DNA and the distance between them is approximately known.

Then occurs the pairwise comparison of all the fragments to estimate the pairwise relationship. Each fragment sequence is compared with the other similar fragment and sequencing and similarity are estimated. The next step is to cluster the fragments which share similarities. Thus, clusters are formed and are used to form data lay-out structures to align the maximum spanning sequences in a linear fashion.

Steps in the genome assembly process

1. Conversion of data from automated sequencer to nucleotide sequence.
2. Removal of unwanted sequences.
3. Assembly of DNA fragments using the overlap sequence similarity.
4. Assessment of assembly process.

Step 1: Conversion data from automated sequencer to nucleotide sequence

The gel image of the automated sequencer is usually a chromatogram, by using computer software; each peak is converted to a particular base. This process is known as **base calling**. Usually the automated sequencers are equipped with this software. But for assembly purpose, the sequence quality is very important. By using some other software like Getlanes or GelImager, once again the automated sequencer data is checked for base calling. This provides an enhanced quality nucleotide sequence. The chromatogram obtained from the semi-automated or automated sequencers has areas of good quality and of poor quality. This has to be examined by the user. There are computer programs for base callers such as Phred or BaseFinder that perform the automated checks of the traces and provide good quality nucleotide sequences for the assembly. These programs allot a specific value to each base which denotes the accuracy of the base.

Step 2: Removal of unwanted sequences

The DNA fragments are not used for sequencing directly after isolation from a particular genome. First, they are cloned in **plasmid** vectors and multiplied in *E. coli* host. Using vector **primers**, they are sequenced from both ends. Therefore, the vector sequences represented in the chromatogram is not related to the original genome. They must be removed from the sequence file before it is submitted to the assembly program. These vector sequences are identified and removed by the base caller computer software program. This process is known as trimming of the sequence. Sometimes even the host bacterial DNA or organelle DNA might be sequenced so as to be identified and removed. Some of the base caller software programs mask the vector sequence during the assembly step.

Step 3: Assembly of DNA fragments using the overlap sequence similarity

The assembly process is carried out by a specialized computer program called an assembler. The input to the assembler program is usually a raw sequence of overlapping randomly fragmented DNA clone sequences. The output from the assembler is the contigs. The assembly process is affected by many factors such as error rate and presence of repeat sequences. The assembler program performs pairwise comparison of all the DNA sequence of fragments, and similarities of all overlapping sequence sequences are identified. A threshold value is set for the similarity, based on which it is decided whether to include the similarity of the two sequences or to discard them. The overlapping sequences are assembled to form clusters. A consensus sequence is identified to form a contig. Initially, the number of contigs increases as the overlapping DNA fragment sequences are joined to form contigs, later some of the contigs sharing the same overlapping sequences are joined. Subsequently, the contig number decreases as the number of new overlapping DNA fragment sequences is added to the already existing contigs. If the contig is prepared using BAC clones, then it is called a **Bactig**. This shows that the coverage of the sequence is good. Mostly the assembler will not give perfectly reconstructed genomic DNA, instead it gives only contigs. This results in gaps in the assembled genome. The gaps in the assembled genome occur for many reasons which are discussed in the next section. Another program is called the scaffolder, which orders and orients the contigs into a longer DNA sequence called a supercontig, in order to form the original large genome. The first sequence assembler which was built for large genome assembly was tested on the *Drosophila* genome sequence.

Step 4: Assessment of assembly

Once the assembly has been prepared, it has to be subjected to editing to assess the quality of the assembled sequence. The most commonly used editing software is consed. The distance between end sequences of the clone gives some indication of the assembly. If the distance between them is approximately equal to the clone insert length, it indicates the correct order of assembly, otherwise there is potential misalignment. For a prokaryotic genome, which is a single large structure, the assembler program should produce one single large assembled sequence. In the case of eukaryotes, it has to produce one large sequence for each chromosome. But this does not always happen when the fragmented genome is reconstructed using the assembler program. However, supercontigs are produced with many gaps between them. The non-contiguous DNA is produced by the assembler for many reasons, such as errors in sequencing, or lack of representation of a particular region of the genomic DNA in the library. Sometimes the contigs are misassembled due to the presence of the repeat sequence in many different places.

8.3 Genome assembly program

The basic requirement for the assembly algorithm is that every sequence read and presented to the algorithm must be there in the genome. The sequence obtained from the automated sequencer is highly fragmented and they are not connected. The assembler is a computer algorithm which performs the sequence assembly based on overlapping sequences. The assembler program joins the overlapping DNA reads.

8.3.1 PHRAP

PHRAP is the PHRagement Assembly Program and it is one of the earliest programs developed for genome assembly. Although it was developed to assemble shotgun sequence reads, it is also capable of assembling the EST sequences and DNA polymorphic sequences. It uses the entire genome sequence to assemble and it does not require high quality data. It also works well with shotgun sequences with vector sequence and low quality traces. It was first used to assemble the shotgun genome sequence assembly of C. elegans and then it was used to draft the sequence assembly of the human genome.

8.3.2 CAP3

CAP 3 is Contig Assembly Program 3. When raw shotgun sequence data from an automated sequencer is fed to this assembler program, it first identifies and trims the vector sequences and also generates the low quality 5' and 3' traces. Then it looks for the complete identical sequences without gaps between the shotgun sequence DNA fragments. The overlapping sequences are assembled using the Smith–Waterman algorithm to calculate the similarity scores for each overlap. Then the overlapping DNA fragments are assembled and the consensus sequence is deduced from multiple sequence alignment of the overlapping sequences.

8.3.3 Overlap lay-out consensus

This uses graph theory in which the reads are represented as edges, and if the overlapping edges share similarities that are connected by nodes the assembly problem is solved by joining the nodes. Initially, based on the overlapping nodes, a graph is constructed. This is known as the lay-out. At this stage, redundant connects are removed. Graph algorithms are used to find out the consensus sequence of the original genome sequence.

8.3.4 Eulerian path

The Eulerian path is based on the sequence by hybridization approach. In this method, a set of sequences of length of K-mers is identified throughout the genome. Each sequence read is broken down into overlapping K-mers. Overlapping K-mers are connected and represented in the graph. The assembly process is completed by the Eulerian path in which all the edges are connected. In this approach, repeats are identified immediately.

8.4 Gaps and gap closing methods

Ideally, the ultimate aim of the genome sequencing projects is to achieve the complete sequence of the original genome. For genome sequencing, many copies of the genome to be sequenced are fragmented and redundant clones are randomly selected for sequencing. Gaps are inevitable in any genome sequencing techniques. After assembling the randomly fragmented sequences into contigs, some gaps do exist. There are two types of gaps: physical gaps and sequence gaps. This is due to two reasons: a particular clone may not be picked up during sequencing or a particular DNA is not present in the library.

8.4.1 Gap closing methods

The gaps can be closed using well-planned, directed sequencing strategies. Usually these gap closing techniques are time-consuming and also involve costlier steps in the whole genome sequencing. They also do not align different-sized genomes. On the other hand, an algorithm developed, known as Projector-2, can make alignment between two genomes with a 40 per cent size difference. The added advantage of the Projector-2 software is that it can design the primers automatically. Repeat masking is another feature of this program in which the given input sequence is scanned for the repeats and masked before assembly.

Linking clone strategy

The linkage clones of the unlinked contigs can be obtained from the genomic sequences of the related organisms. Some software packages (MuMmer, MGview) were developed to align the genomes and identify the linking contigs. These software packages do not provide automatic primer designing to find the linking clones. Manual examination is important to identify the linking clone sequences.

Primer walking

Primer walking is another strategy adopted to close the gaps, but this requires sequential hybridization experiments to find out the linking clone. Gaps of different sizes are located at different areas of the assembled genome sequence. The cost of closing these gaps varies with the type of strategy used for sequencing.

PCR amplification

The easiest method of gap closing is PCR amplification of the library of clones with unique primers designed from the ends of the contigs. Practically, it can rarely be done, because it is very difficult to get unique primers from the ends of the contigs with less than 10 kbp length. Often the gap size is more than10 kbp; hence, it becomes difficult to close the gap with the PCR amplification strategy.

8.5 Draft and finished genome sequences

The genome of an organism, especially higher eukaryotes like a human being, consists of a euchromatic region (expressed part of the genome in which most of the genes are located) and a heterochromatic region (highly condensed **heterochromatin** which is present in the **centromere** and **telomere** regions). When genome sequencing projects started in 1990, the main goal of the genome project was to know the number of genes present in an organism sequenced. Therefore, initial genome sequencing efforts were made to sequence the euchromatic regions of the chromosome.

8.5.1 Draft genome sequence

As soon as the euchromatic regions were sequenced, they were assembled to provide a 'draft genome sequence'. The *draft genome sequence* is characterized by the presence of gaps, i.e. the genomic DNA is represented as supercontigs rather than single chromosomes, with the presence of base ambiguities and low accuracy; in other words, the presence of errors in the sequence and misalignment in the ordering of contigs. Although the draft sequence has many pitfalls, it provided vital data for further progress of the genome sequencing projects. It provided a comprehensive estimate of the number of genes present in a particular genome. For example, initially it was predicted that a human genome might contain over 80,000–1,00,000 genes. After the draft, the human genome sequence was published in 2000; the annotation process showed that it actually contains only 20,000–25,000 genes. Another aim of the genome sequencing process is annotation of the genome sequence in terms of genes and associated elements.

The draft genome sequence has gaps, ambiguities, less **depth of coverage** (the number of times a particular part of genome is being sequenced is known as depth of coverage). It denotes the accuracy of the sequence being determined at that particular position. Usually a particular nucleotide is determined, based on the consensus of the assembled reads. The finished sequence will have minimum gaps and the depth of coverage is at the more

Published by Woodhead Publishing Limited

usually accepted level of 8X to 10X. The error rate is less: the internationally accepted level is one error in 10,000 bases.

8.5.2 Finished genome sequence

The next stage of the genome projects was to prepare the *'finished genome sequence'*. The gaps were closed by using a targeted sequencing strategy; base ambiguities were removed and the bases conformed by sequencing more number of clones, i.e. the depth of coverage was increased from 4X in the draft genome sequence to 10X in the finished genome sequence. Therefore, the finished genome sequence is characterized by no gaps, ordered clone contigs, accurate base calling; the error rate is extremely low and the accepted error is 1/10000. The contigs were more correctly ordered by comparing them with high density physical maps. Complete reconstruction of the original genome that represents the whole genome sequence in a single contig was achieved only in bacterial genome sequences and some lower multicellular eukaryotes. The finished genome sequence can be used for comparative genomics because the genome sequence is complete and ordered. The exact location of the genes and their associated elements can be determined for higher eukaryotes, as their genomes have highly repetitive sequences in centromeres and telomeres. Preparing a finished genome sequence for higher eukaryotes is a terrible task as their genomes have highly repetitive sequences in the telomere and centromere regions. It is also a laborious, time consuming and costly affair, although it is the ultimate objective of any genome sequencing project. Table 8.1 gives a brief comparison between the draft and the finished genome sequence.

Table 8.1 Comparison of draft genome sequence and finished genome sequence

Sl. No.	Draft genome sequence	Finished genome sequence
1	Incomplete genome sequence	Near complete genome sequence
	Characterized by presence of gaps, base ambiguities and low depth of coverage	Characterized by no gaps, depth of coverage is high and no base ambiguities
2	Cannot be used for comparative genomics study	Can be used for comparative genomics study
3	Comprehensive estimate of genes present in the organism	Describes the location of genes and their associated elements
4	Error rate is high	Error rate is low, internationally accepted error rate for finished genome sequence is 1/10,000

Published by Woodhead Publishing Limited

8.6 Genome annotation

The genome of all organisms is made up of a sequence of nucleotides which consists of four different bases. The genome sequencing projects deduce the order of the bases in the genome. But to understand the genome and its function, the sequencing has to be further characterized in terms of the genes and their associated regulatory elements. Genomes of higher organisms consist of many repeat sequences and their function is not known. Therefore, the genome sequence has to be classified and its function has to be assigned. This process is known as genome annotation. It is usually done by automated annotation computer programs although manual curation is also done. It is a bioinformatics work. The basic level of annotation is done using BLAST to find out similarities.

Gene prediction in eukaryotic genome sequence is challenging due to its huge size, the occurrence of repetitive sequences, the low gene density and split genes. On the other hand, gene prediction in prokaryotic genomes is straightforward. Gene prediction programs can be grouped into two categories: (1) *ab initio* programs which predict genes from a DNA sequence based on nucleotides such as start codon (ATG) and stop codon (UAA or UGA or UGA); and (2) homology-based gene prediction programs which predict genes based on percentage sequence similarity with a known gene sequence. The most commonly used computer program to predict genes and related regulatory elements from the genome sequence is GRAIL (Gene Recognition and Assembly Internet Link) developed by Dr Ed Uberbacher at Oak Ridge National Laboratory, and supported by the DOE.

Gene prediction programs: *Ab initio* methods

1. Fgenesh
2. Glimmer
3. GlimmerM
4. Genemark
5. GRAIL

8.7 Comparative genomics

Initially organisms were grouped according to their morphological features. Protein sequences and DNA sequences are used as tools to compare their evolutionary relatedness in phylogenetic tree construction. Phylogenetic trees are the most reliable classification method to study the organisms' relations. Bursts of genetic information in the genomics and proteomics era paved the way for comparison to be made between every organism on Earth based on the protein and DNA sequence comparison. Previously it was not possible to make a comparison between a bacteria and a human but it is now possible.

Goals of comparative genomics

1. To identify orthologous and paralous genes in different species.
2. To predict the most recent ancestor for each species.
3. To determine the positively and negatively selected genes in different species.
4. To identify the core genome of each species.
5. To know about the births and deaths of genes.

Cataloguing the genes and their function is the next stage of the genome project. In comparative genomics, sequence alignment is the core process. An alignment is a process in which the nucleotide sequence of one genome is matched against another; the number of matching positions can be increased by introducing gaps in the genome sequence. Comparison of the genome sequence is not performed manually; instead many alignment algorithms are developed to do so. Sequence alignment can be done pairwise in which two genome sequences are aligned. More than two genome sequences can also be aligned which is called multiple sequence alignment.

Comparative genomics deals with the structure and organization of the genome and their order across living organisms. The basis for comparative genomics is that in all living organisms DNA is the genetic material, and the building blocks of the DNA are common. This branch of genomics is a newly developed field which relies on the complete genome sequence data of the organisms to be compared. The main requirement for comparative genomics is vast genome sequence data and computer programs. Comparative genomics is useful in describing a genome in terms of gene content, repeat content, gene order conservation, etc. The presence and absence of genes make the difference between organisms.

Predicted functions of the genes can be assigned with some degree of confidence using conservation of that gene across organisms. Analysis of genomes of bacteria has identified that there is bias towards accumulation of mutation in the strand and the presence of genes in one of the strands.

From comparative studies on the bacterial genomes, it has been identified that there is correlation between arrangements of the gene from the origin of replication to the terminus. Genes located nearer to the terminus have low G+C content and show faster evolution rate. Pan genome is the total number of genes present in a species. Core genome denotes the total number of genes present in a monophyletic group. It has been shown that all living organisms on Earth share very few genes and it has been estimated that 60 genes are present in all of them. Comparative genomics provides more insight into the evolutionary relationships, biochemical pathways, gene expression patterns, gene order, etc. that can be identified.

Comparative genome alignment programs

1. DIALIGN: pairwise and multiple sequence alignment
2. MUMmer
3. Vmatch
4. WABA success
5. PipMaker
6. VISTA/AVID

Review questions and answers

1. *What is genome annotation?*

The genome sequence has to be named and its function has to be assigned. This process is known as genome annotation.

2. *What is the draft genome sequence?*

The draft genome sequence is characterized by the presence of gaps, i.e. the genomic DNA is represented as supercontigs rather than single chromosomes, with the presence of base ambiguities and low accuracy, otherwise presence of error in the sequence, misalignment in ordering of contigs.

3. *Why are there gaps in the genome assembly?*

There are two types of gaps such as the physical gap and the sequence gap. This is due to two reasons: a particular clone may not be picked up in sequencing or a particular DNA is not present in the library.

Published by Woodhead Publishing Limited

4. *What is a contig or Bactig?*

A contig is the assembly of overlapping clones without a gap, i.e. the unbroken series of clones assembled using overlapping sequences. Bactigs are contigs prepared from BAC clones.

Recommended reading

Abby, S. and Daubin, V. (2007) 'Comparative genomics and the evolution of prokaryotes', *Trends in Microbiology*, 15: 135–41.
Cai, W.W., Chen, R., Gibbs, R.A. and Bradley, A. (2001) 'A clone-array pooled shot-gun strategy for sequencing large genomes', *Genome Research*, 11: 1619–23.
Frangeul, L., Nelson, K.E., Buchrieser, C., Danchin, A., Glaser, P. and Kunst, F. (1999) 'Cloning and assembly strategies in microbial genome projects', *Microbiology*, 145: 2625–34.
Huag, X. and Madan, A. (1999) 'CAP$_3$: a DNA sequence assembly program', *Genome Research*, 9: 868–77.
Koonin, E.V. (2003) 'Comparative genomics, minimal gene-sets and the last universal common ancestor', *Nature Review of Microbiology*, 1: 126–36.
Lapuk, A., Volik, S., Vinsent, R., Chin, K., Kuo, W.L., de Jong, P., Collins, C. and Gray, J.W. (2004) 'Computational BAC clone contig assembly for comprehensive genome analysis', *Genes, Chromosomes and Cancer*, 40: 66–71.
Mardis, E., McPherson, J., Martienssen, R., Wilson, R.K. and McCombie, W.R. (2002) 'What is finished and why does it matter?', *Genome Research*, 12: 669–71.
Roach, J.C., Slegel, A.F., den Engh, G.V., Trask, B. and Hood, L. (1999) 'Gaps in the Human Genome Sequencing Project', *Nature*, 401: 843–5.
Rubin, G.M. et al. (2000) 'Comparative genomics of the eukaryote', *Science*, 287: 2204–15.
Sacha van Hijum, A.F.T., Zomer, O.S. and Kok, J. (2005) 'Projector-2: contig mapping for efficient gap-closure of prokaryotic genome sequence assemblies', *Nucleic Acid Research*, 33.
Schmutz, J. et al. (2004) 'Quality assessment of the human genome sequence', *Nature*, 429: 365–8.
Staden, R. (1980) 'A new computer method for the storage and manipulation of DNA gel reading data', *Nucleic Acid Research*, 8: 3673–94.
Wang, Z., Chen, Y. and Li, Y. (2004) 'A brief review of computational gene prediction', *Geno. Prot. Bioibfo*, 4: 216–21.

Web addresses

http://combio.ornl.gov/grailexp/.
http://genometools.org/.

Published by Woodhead Publishing Limited

9

Functional genomics

Sequencing the genome . . . is only the beginning of genomics.

Neal Lane, Director of the Office of Science
and Technology Policy, USA

Abstract: Living organisms delicately regulate their body functions by carefully modifying the gene expression pattern. Many methods are available to study expression patterns in cells. After the genome sequencing, much attention has been paid to developing gene expression studies. A separate branch of genomics called functional genomics has developed. In this chapter, different expression profiling methods are discussed. We limit our discussion to understanding the functions of genomes from a transcriptome, although functional genomics includes proteomics, phenomics and metabolomics.

Key words: DDRT-PCR, expression profiling, forward genetics, microarray, microSAGE, Northern blotting, reverse genetics, SAGE, SH, SSH.

Key concepts

- All the genes are not expressed at all times; few genes are expressed all the time while some genes are expressed only sometimes.
- Earlier experiments were designed to study the gene expression pattern of a single gene.
- High throughput gene expression methods involve techniques which study all transcripts or proteins expressed in a cell or a tissue.
- Northern blotting is the oldest gene expression analysis method which studies one gene at a time and provides information on relative expression levels of mRNA.

Published by Woodhead Publishing Limited

- DDRT-PCR and representational display analysis (RDA) are PCR-based methods that are used to study the differential expression of genes at the mRNA level.
- SAGE is both a qualitative and a quantitative high throughput sequencing-based method to study all the transcripts that are expressed in a cell or a tissue.
- Microarray is a hybridization-based high throughput transcript analysis method which is capable of analyzing the expression pattern of thousands of genes simultaneously.
- High throughput gene expression methods like SAGE and microarray require computer software for data analysis.

9.1 Introduction

To know the cellular functions of any organism, understanding its gene function is very important. Traditionally molecular biologists studied the gene function one at a time. There were no gene sequence and corresponding high throughput gene expression analysis techniques. Now, the complete genome sequence of many important organisms is available; the time has come to study many genes in a single experiment. The ultimate goal of the genome sequencing project is to know the function of all the genes present in a genome. The information which can be assigned to any gene expression status could be: (1) when the gene is expressed; (2) how much of it is expressed; (3) the parts in which the particular gene is expressed; and (4) what the other gene products are with which the particular gene interacts.

Traditional gene function analysis involved knowing the phenotypes of the gene first and then finding the gene and its sequence. This is popularly known as **forward genetics**. Finding the functions of the gene from its sequence is known as **reverse genetics**. After sequencing the whole genome of an organism, one might ask, what do we do next? Merely depositing millions of DNA sequences in the database will not yield any clue about the functions of the genes and its related sequences. Here's where **functional genomics** comes in. It helps to predict the functions of the genes from their sequences using both the wet lab and dry lab methods. Unless the function of a gene is explained, the genome sequence data lacks its ultimate purpose, and enormous amounts of money have been spent in vain. Therefore, functional genomics adds value to the genomics data and ultimately makes them useful for human welfare.

Functional genomics involves the use of high throughput techniques to study all the genes, all transcripts and all proteins expressed at a particular time in a tissue. It describes the functions of the genes and their interactions. Different methods are used to study the gene expression. Traditional methods were based on the hybridization principle and they give information about a single gene function. After the genomics era, many gene expression profiling methods were developed which are characterized by automation, analyzing multiple samples, the requirement of a very low quantity of starting material, and the simultaneous analysis of thousands of genes.

9.2 Northern blotting

Three decades ago, the expression of genes at the RNA level was studied using the **Northern blotting** technique and it was one of the first few techniques developed to study gene expression at the transcript level. It was described by Alwins et al. in 1977. The name 'Northern blotting' was given to this technique as the procedure used is analogous to the **Southern blotting** technique. Northern blotting involves separation of RNA in agarose gel under denaturing conditions and the transfer of the RNA separated on agarose gel to the nylon membrane using capillary transfer. The specific transcript is detected using a labelled probe which is complementary to the transcript. Using this technique, the expression status of a single gene can be studied. The entire Northern blotting procedure can be divided into three steps (Figure 9.1).

Steps in Northern blotting

1. Preparation of RNA sample.
2. Separation of RNA sample in formaldehyde agarose gel and transfer of the RNA from gel to nylon membrane.
3. Hybridization with labelled probe and detection.

Step 1: Preparation of RNA sample

Purified mRNA is preferred for Northern blotting. From the tissue sample, the total RNA is extracted and then mRNA is purified using a separate technique called oligo (dT) affinity chromatography which exploits the poly(A) tail of the eukaryotic mRNA.

Step 2: Separation of RNA sample in formaldehyde agarose gel and transferring the RNA from gel to nylon membrane

The quality and quantity of the RNAs can be assessed after separating them in agarose gel. RNAs are single-stranded nucleic acid molecules and they tend to form secondary structures through intramolecular hydrogen bonding which affects the mobility of the RNA during electrophoresis. For effective probe hybridization, target mRNA molecules have to be maintained in a single-stranded form. Therefore, electrophoresis is carried out in the presence of denaturing agents like urea and formamide which remove secondary structures. A good quality total RNA preparation will show two bands; the upper band corresponds to 28S and the lower band 18S rRNA. As the separated mRNA target molecules are located inside the gel matrix, they cannot be accessed for hybridization, as probes cannot penetrate the agarose gel matrix. Therefore, mRNA molecules that are separated are transferred to a solid support which can easily be accessed by the probes for hybridization. A positively charged nylon membrane is used for Northern blotting. A nylon membrane has positively charged primary amines which enable ionic bonding with the negatively charged phosphate group of nucleic acids and helps the firm attachment of RNA to the membrane.

Step 3: Hybridization with labelled probe and detection

The mRNA molecules immobilized on the nylon membrane are hybridized with the labelled probe. cDNA or genomic DNA molecules with sufficient homology are used as probe. The hybridization buffer often contains formamide which prevents degradation of mRNA at the high temperature maintained during hybridization, and it also lowers the annealing temperature allowing one to work at low temperature. After hybridization the membrane is thoroughly washed to remove the excess probes under highly stringent conditions. Under highly stringent conditions, only the target-probe hybrid is retained on the membrane and other unbound labelled probes are removed. The location of the target mRNA molecules on the membrane can be detected by autoradiogram. BlotBase is an online database source for Northern blots of humans and mice. It can be accessed using their blot ID, gene identifier and article reference.

Published by Woodhead Publishing Limited

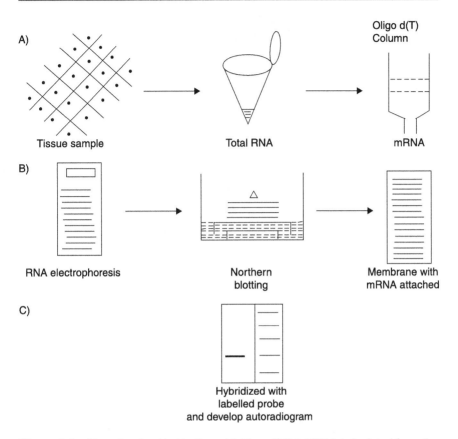

Figure 9.1 Steps involved in Northern blotting. A) Total RNA is isolated from the sample and mRNA is purified by oligo (dT) affinity chromatography. B) mRNA is separated in agarose gel and transferred to a nylon membrane by capillary transfer. C) The nylon membrane with mRNA is hybridized with a labelled probe and the autoradiogram is developed to locate the specific mRNA transcript.

Applications of Northern blotting

1. It is used to identify the expression status of the gene.
2. It is also possible to quantify the amount of transcript in a particular tissue.
3. It is possible to compare the expression status of the two related tissues, for example, healthy tissue vs cancerous tissue.
4. mRNA size can be determined.
5. Alternate splice products can be identified.

Published by Woodhead Publishing Limited

Limitations of Northern blotting

1. Only a single RNA molecule or only a single gene expression is studied at a time.
2. Handling RNA throughout the experiment is difficult as it is very prone to degradation.

Summary

Although Northern blotting is a good technique to study the gene expression, due to its laborious procedure and the limitation of studying only one gene at a time, it is less appealing in the light of other **transcriptomics** techniques. Northern blotting can also be used to validate the genes identified using transcriptomics techniques such as DDRT-PCR, SAGE and microarray.

9.3 Subtractive hybridization

All genes are not expressed at any given time in all the tissues. Some genes are expressed specifically in some tissues only; they are called tissue-specific *genes*. It is crucial to understand the expression of tissue-specific genes to find out their role. For example, some genes are over-expressed only under pathological states. If we could identify and isolate those gene products, one could probably identify new drug targets for treatment or biomarkers for diagnosis. Subtractive hybridization was the first technique developed to study the tissue-specific gene expression.

9.3.1 Principles of subtractive hybridization

This uses the mRNA pool of one sample which is called a driver and cDNA pools of other samples which is called a tester. The driver mRNA pool has many of the mRNA in common with the tester. Tissue-specific cDNA is subtracted from the cDNA pool of the sample by hybridizing with the driver mRNA population. When the tester (biotinylated cDNA) and the driver (excess amount of mRNA) pools are mixed, the common sequence between tester and driver are hybridized to form double-stranded cDNA and mRNA and the single-stranded cDNA which are left are separated. The subtracted cDNA is cloned and sequenced to identify the genes (Figure 9.2).

Published by Woodhead Publishing Limited

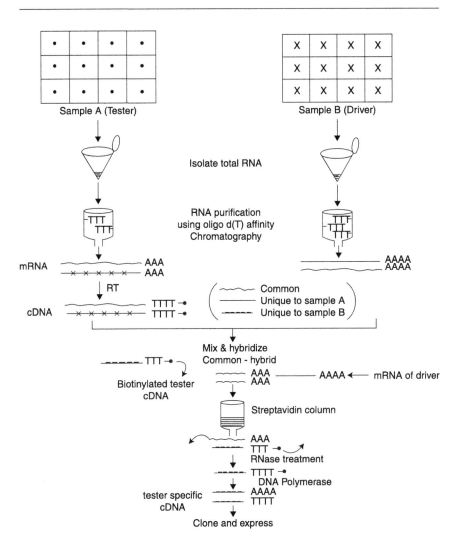

Figure 9.2 Steps involved in subtractive hybridization. mRNA is purified using oligo (dT) affinity chromatography. Tester mRNA is converted to cDNA by reverse transcription using biotinylated oligo (dT) primer. A tester cDNA pool is allowed to hybridize with an excess amount of driver mRNA pool. Under appropriate conditions, double-stranded cDNA–mRNA is formed by common sequences and the tissue-specific tester cDNA and the driver mRNA are maintained as a single-stranded molecule. Biotinylated single-stranded cDNA is separated using streptavidin column.

Steps in subtractive hybridization

1. Preparation of tester mRNA and driver mRNA pools.
2. Preparation of tester cDNA using biotinylated oligo (dT) primer.
3. Hybridization of labelled cDNA of tester and driver mRNA pools.
4. Subtraction of tester cDNA from cDNA/mRNA hybrids.
5. Cloning and sequencing of the subtracted cDNA.

Initially the **subtractive hybridization (SH)** method that was developed relied on hydroxyapatite column to separate the single-stranded and double-stranded hybrids, but this is time-consuming and also complete separation is not achieved. Later, biotin labelling of tester cDNA was developed which helped to purify the subtracted tester cDNA using biotin-strepdavidin affinity chromatography. To increase the sensitivity of SH, subtracted tester cDNA is PCR amplified over the common mRNA molecules. This completely eliminates the separation step. This is known as suppression subtractive hybridization (SSH). Now there are commercial kits available to perform SSH.

Applications of subtractive hybridization

1. Useful in finding probes for cDNA library screening.
2. Tissue-specific cDNA clones can be isolated for cloning and expression.

Advantages of subtractive hybridization

1. It does not require prior knowledge about the transcript.
2. No specific high cost equipment is needed.

Disadvantages of subtractive hybridization

1. It requires a large quantity of mRNA which is difficult to prepare.
2. It is applicable only for pair-wise comparison.

Published by Woodhead Publishing Limited

3. It does not provide a quantitative measure of the gene expression.

Summary

SH is a hybridization-based gene expression analysis method. Using this technique, tissue-specific genes can be isolated. Although it helps to isolate many tissue-specific genes from a single experiment, its use in functional genomics study is limited by its inherent requirement of comparative mode of study and lengthy tedious steps involved.

9.4 Differential Display Reverse Transcription PCR (DDRT-PCR)

DDRT-PCR is considered as an alternative to the laborious Northern blotting technique to study the gene expression. It is a PCR-based method and is capable of detecting the expression of many genes at a time. It works better under gene expression studies in a comparative mode, for example, gene expression analysis in healthy vs diseased, control vs treated, etc. In the DDRT-PCR method, mRNA is converted to cDNA by reverse transcriptase enzyme using an oligo (dT) primer. Subsequently, 3' ends of mRNA population are amplified using anchored oligo (dT) primer and random primer. The PCR amplified subsets of cDNA fragments are separated in a high percentage of polyacrylamide gel. Two related samples are run side by side. By comparing the banding pattern, one can detect the common transcripts and specific transcripts of a particular tissue type. Hence, the name of the technique is **differential display reverse transcription** (DDRT-PCR). Although initial studies on DDRT-PCR were done using radioactive isotopes to detect the PCR amplified products on the sequencing gels, many other non-radioactive methods were also developed, including using silver staining and chemi-luminescent assay. Owing to its simplicity, it has become a handy tool to study tissue-specific gene expression. Figure 9.3 presents the steps involved in DDRT-PCR.

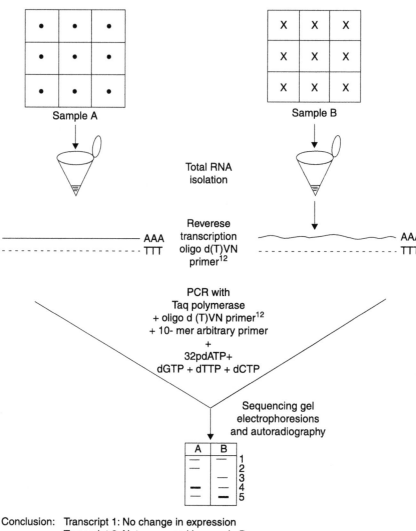

Conclusion: Transcript 1: No change in expression
Transcript 2: Not expressed in sample B
Transcript 3: Not expressed in sample A
Transcript 4: Over-expressed in sample A
Transcript 5: Over-expressed in sample   B

Figure 9.3   Steps involved in DDRT-PCR. Total RNA is isolated from the samples and mRNA is purified using oligo (dT) affinity chromatography. cDNA is synthesized using reverse transcriptase using oligo (dT) primer. Subset of cDNA is PCR amplified using anchored oligo (dT) primer and a random primer in the presence of radioactively labelled dNTPs. The PCR amplified products are separated using sequencing gel and an autoradiogram is developed. Based on banding position and intensity, a conclusion is made on the gene expression status between samples.

Steps involved in DDRT-PCR

1. The total RNA is extracted from the selected tissues and mRNA is separated using the oligo (dT) column.
2. Synthesis of cDNA by reverse transcription of mRNA using oligo (dT) to anchor primer.
3. Synthesis of second strand using oligo (dT) primer and primer in the presence of labelled dNTPs.
4. PCR amplification of cDNA using anchored oligo (dT) primer and a random primer.
5. Separation of PCR amplified products in sequencing gel.
6. Exposure to X-ray film and development of an autoradiogram.

Applications of DDRT-PCR

1. It enables one to develop probes for tissue-specific genes.
2. It helps to clone tissue-specific genes and enable their further characterization.

Advantages of DDRT-PCR

1. It is a sensitive technique.
2. No special equipment is needed.
3. Expression of many genes can be studied in a particular experiment.
4. No prior information about the mRNA is needed.

Disadvantages of DDRT-PCR

1. Presence of many false positive results.
2. Since DDRT-PCR products are truncated, they cannot be cloned and studied.
3. Not useful in single mode as the candidate gene is identified based on comparative expression mode.
4. Although it gives information about the expression status of many genes, it is still insufficient to cover all the genes expressed in a tissue.

Published by Woodhead Publishing Limited

Summary

DDRT-PCR is a simple and sensitive technique to study expression under comparative mode. False positive results from this technique prohibit the use of this technique. The main reasons for false positives are the binding of primers to contaminated DNA, or the binding of primers to a non-specific sequence at low annealing temperature. To overcome this, DNA contamination in mRNA sample is removed by treating the mRNA sample with DNase prior to reverse transcription.

9.5 Representational Difference Analysis (RDA)

RDA is a PCR-based subtractive hybridization technique for gene expression studies. It helps to isolate tissue-specific genes. It is a high throughput technique for gene expression analysis and it does not need any prior information on cDNA sample. It was originally developed to study differences in genome sequences. RDA involves amplification and cloning of a unique cDNA population from a sample by hybridization between the tester and the driver. It involves two steps: (1) subtraction of the tester homoduplex from the driver; and (2) PCR amplification of tester homoduplex. To enrich the tester homoduplex, multiple rounds of subtraction are performed by increasing the tester: driver ratio from 1:100, 1:800 and 1:40,000 in the first, second and third cycles of subtraction respectively. cDNA from two samples are restricted with DpnII enzyme (four base-cutting type II enzyme) and produce four base pair 5' overhangs. The restricted cDNA are ligated to linkers. Using primers specific to linker molecules, PCR is carried out. Each mRNA in the original sample is represented as cDNA amplicon. Therefore, cDNA amplicons are called representations.

Steps in RDA

1. Isolation of total RNA and purification of mRNA.
2. Synthesis of cDNA.
3. Generation of representations after digesting with DpnII enzyme.
4. Hybridization of tester and driver cDNA presentations.
5. PCR amplifications of unique representations.
6. Cloning and sequencing of the different cDNA products.

Step 1: Preparation of mRNA samples

Total RNA is isolated from tester and driver samples. Total RNA used for RDA should not have contaminants like DNA. A sufficient quantity of high quality RNA should be isolated as it is required for both RDA and Northern blotting. Northern blotting is done to verify the cDNA representations that are identified using RDA. Therefore, 500 μg of the total RNA is required. DNA contamination in the RNA preparation is removed by treating the total RNA with RNAse-free DNAse. The mRNA population is purified from the total RNA isolated using oligo (dT) affinity column chromatography. Approximately 5–10 μg of poly (A) tail RNA is purified from 250 μg total RNA. Since purified mRNA is labile and easily degraded, it should be used immediately for cDNA synthesis.

Step 2: Synthesis of cDNA

The first strand of the cDNA is synthesized using oligo (dT) as primer by using reverse transcriptase. The second cDNA synthesis is carried out using DNA polymerase. A number of cDNA synthesis kits are available from various companies. After cDNA synthesis, the quality of the cDNA should be analyzed using 1 per cent agarose gel. A good quality cDNA preparation shows an intense smear between 200 and 300 bp.

Step 3: Generation of representations after digesting with DpnII enzyme

cDNA representations are prepared from double-stranded cDNA. Approximately 2 μg cDNA is subjected to DpnII digestion. After phenol-chloroform extraction, cDNA is precipitated with ethanol. The cDNA fragments are ligated to adapter and PCR amplification is performed with these primers. Each cDNA in the original sample is amplified and represented in the PCR mixture. These are known as representations.

Step 4: Hybridization of tester and driver cDNA presentations

J-BglI ligated tester cDNA presentations are mixed with driver cDNA presentations to form three hybrids: (1) tester homoduplex; (2) tester:driver heteroduplex; and (3) driver homoduplex.

Step 5: PCR amplifications of unique representations

Using J-BglI adaptor 24 mer primers, PCR amplification is performed. Tester homoduplex is exponentially amplified, tester:driver heteroduplex is amplified linearly, whereas driver homoduplex is not amplified because of lack of primer binding site. This type of hybridization and selective amplification is performed three times in different tester:driver ratios in order to enrich the tester-specific cDNA representations.

Step 6: Cloning and sequencing of the different cDNA products

The final PCR product is cloned into a cloning vector and sequenced. From the sequence, the gene is identified from the database search of the gene which is isolated from the cDNA library for further characterization.

9.6 Serial Analysis Gene Expression (SAGE)

SAGE is a sequencing-based gene expression analysis method, a true high throughput transcriptomics method. It has been shown that a small sequence of the cDNA known as the 'SAGE tag' is sufficient to identify the gene.

9.6.1 Principles of SAGE

A short sequence (approximately nine bases) tag is isolated from 3' end of each mRNA transcript expressed in a tissue by endonuclease digestion of the cDNA population. The multiple tags produced are ligated together to form a contig, PCR amplified and then cloned into a vector for sequencing. After sequencing, the number of unique tags is counted to identify the total number of different genes expressed in that particular tissue. The quantitative estimate of each gene expressed could be theoretically calculated using the relative abundance of the tags. Genes corresponding to the SAGE tag are identified by searching against the cDNA or EST databases.

Steps in SAGE

1. Isolation of total RNA and purification of mRNA.
2. Reverse transcription of mRNA to biotinylated double-stranded cDNA.
3. Generation of tags using tagging enzyme.
4. Dividing the tags and ligating to linker molecules.
5. Preparation of ditags.
6. Digesting with anchoring enzyme.
7. Ligating tags to produce concatenates.
8. Cloning of concatenates and sequencing.
9. Identification of genes and EST and quantification of tags.

Step 1: Reparation of total RNA

The biological tissue sample is homogenized in the RNA isolation buffer. Commercial total RNA isolation kits are available. The quality of the total RNA isolated can be checked in denaturing agarose gel.

Step 2: Purification of poly (A) tail RNA and checking the quality of mRNA

Apart from poly (A) tail mRNA, total RNA has other RNAs like rRNA, tRNA, micorRNA, etc. It is reported that 1 mg of total RNA contains only 20 μg of poly (A) tail RNA. Purification is done using oligo d (T) affinity chromatography. Before proceeding to the SAGE experiment, the quality of the mRNA prepared should be analyzed using northern hybridization with a known probe.

Step 3: cDNA synthesis

This involves both first strand synthesis and second strand synthesis. The first strand cDNA is synthesized using biotinylated oligo (dT) primer by reverse transcriptase. Second strand cDNA is synthesized from the loop created in the first strand by E. coli DNA polymerase and mRNA is removed using RNAse H.

Step 4: Digestion of biotinylated cDNA with anchoring enzyme

The double-stranded cDNA is digested with NlaIII (type III restriction endonucleases, which makes cut 10–15 nucleotides away from the recognition site). It is found that all mRNAs have recognition sites for this enzyme. Each cDNA is cleaved into many fragments. 9–12 nucleotides attached to the 3' end of the cDNA are called a tag.

Step 5: Purification of anchored biotinylated cDNA

The entire mixture prepared in step 4 is extracted with phenol-chloroform to remove the restriction enzyme and other components. The double-stranded cDNA (tag) is precipitated with ethanol and purified using strepdavidin column. Subsequently after the elution step, biotinylated double-stranded anchored cDNA (tag) is released.

Step 6: Ligation with linkers

The eluted biotinylated double-stranded cDNA (tag) is divided into two pools and ligated with two types of linkers.

Step 7: Releasing tags by the tagging enzyme

Biotinylated double-stranded cDNA (tag) molecules contain a recognition sequence for a type III enzyme (FokI) in their linkers. FokI cuts away from the recognition sequence and it releases the tag, hence it is called the tagging enzyme. This mixture is extracted with a phenol-chloroform mixture to remove enzymes and other impurities. The tag molecules are precipitated with ethanol.

Step 8: Ditag formation and amplification

The tags released from the two pools are ligated to form ditags. PCR amplification of the ditags is performed using linker sequences as primers.

Step 9: Formation of concatenate and sequencing

Ditags are released by digesting them with anchoring enzyme. The released ditags are ligated to form a concatenate which is cloned in cloning vectors and sequenced.

Step 10: SAGE tag sequence analysis

The sequence obtained has many tag sequences. They can be analyzed using computer software to count and identify tags.

9.6.2 MicroSAGE

Initially the SAGE method required a large quantity of mRNA. Sometimes one has to work with a small quantity of biological sample which will yield only a limited quantity of mRNA. Modification of the SAGE protocol requiring less initial mRNA from the sample is called microSAGE. It requires 500-fold less mRNA than the ordinary SAGE experiment. The principles of MicroSAGE experiment lie in multiple rounds of PCR amplification of the tags before they are taken for sequencing. MicroSAGE is a single tube experimental procedure as against many stages in SAGE. MicroSAGE requires only 1 ng of mRNA.

Application of SAGE

1. The expression status of many genes can be deduced from a single SAGE experiment.
2. It is possible to quantify the abundance of each gene expressed in a tissue.

Advantages of SAGE

1. It does not require any prior information about the transcripts.
2. From a single SAGE experiment, expression of many genes can be studied.
3. It also provides a quantitative estimate of the mRNA transcripts.
4. It is a high throughput transcriptomics technique.

Disadvantages of SAGE

1. It requires completely sequenced genomes for the study of tags comparison and gene identification.
2. It requires a high throughput sequencing facility.
3. Low abundant cDNA tags are often missed.
4. Wrong identification of genes due to sequencing error or false matching of the tags with gene sequences.
5. It requires a relatively high amount of initial mRNA (2.5–5.0 µg of poly (A)$^+$ RNA).

Summary

SAGE is a sequencing-based high throughput expression profiling method which gives global gene expression profile of tissue under study. It involves identification and quantification of the cDNA using SAGE tags.

9.7 Microarray technology

This is a hybridization-based method of gene expression analysis. This technique involves immobilizing a large number of probes on a solid support in a small area. It is the only method in which the entire procedure is automated. This is also efficient in identifying differentially expressed genes. From a single experiment, thousands of genes can be studied.

9.7.1 Principles of microarray hybridization

In this method a target is labelled and a probe is unlabelled. This is in contrast to Northern blotting wherein the probe is labelled and the target is

Table 9.1 Differences between Northern blotting and microarray

Northern blotting	Microarray
Target is immobilized	Probe is immobilized
Probe is labelled	Target is labelled
Single gene is analyzed	Thousands of genes are analyzed

unlabelled. Purified mRNA samples are labelled with fluorescent dyes. Labelled targets are hybridized with immobilized cDNA on a microarray chip. After washing, only hybrids are retained and are identified by measuring the fluorescence. The intensity of the fluorescence light indirectly measures the abundance of the transcripts. Table 9.1 shows the differences between Northern blotting and microarray.

Steps in microarray experiment

1. Preparation or procuring microarray chip.
2. mRNA isolation and fluorescent labelling.
3. Hybridization.
4. Image capturing.
5. Data analysis and interpretation.

Step 1: Preparation of microarray

Depending on the source of the probes, there are two types of microarrays: spotted microarray and oligonucleotide microarray or DNA chip. Microarray preparation consists of three steps: (1) probe selection; (2) PCR amplification of probes; and (3) array printing. For a spotted microarray, cDNA are used as probes, whereas, for DNA chips, oligonucleotides are used as probes. PCR amplified cDNA probes are spotted on glass slide using a robotic array machine. cDNA probe of size 500 bp to 5000 bp is PCR amplified and used for spotting. Spotting is done by robotic microarrayer fitted with printing pins. There are two types of printing or spotting: contact and non-contact. Non-contact printing works on the piezoelectric principle. It dispenses a picolitre quantity of DNA. These pins are capable of taking 0.25 µl DNA solution and dispense about 0.6 nl which is equal to 1–10 ng of DNA. Each spot is about 100 µm in diameter. This technique was first tried at Stanford University, USA. The data obtained from the microarray has to be interpreted to get gene expression. There are many systemic errors which lower the accuracy of results. Factors that contribute to the errors are surface chemistry, microarray printing, labelling methods and image capturing. To account for the above variations, the cDNA library is normalized.

Step 2: mRNA isolation and fluorescent labelling

Total RNA is isolated from control and test samples and mRNA is purified. Each mRNA sample is converted to cDNA in the presence of fluorescent dNTPs. One sample is labelled with Cy3 fluorescent dye and other one is labelled with Cy5 fluorescent dye.

Step 3: Hybridization

Pre-hybridization is carried out to convert the double-stranded cDNA into single-stranded molecules for hybridization. Usually it is done at 42°C for 45 minutes. Labelled cDNA from two samples are pre-heated, mixed and flooded on a microarray chip. A specific condition is maintained for probe–target base pairing. Hybridization is carried out for 16–20 hours. Excess target is washed off and the microarray chip is dried.

Step 4: Image capturing

After hybridization, the microarray chip is scanned with a laser light of different excitation wavelengths that corresponds to the excitation wavelength of Cy3 (550 nm) and Cy5 (649 nm) fluorescent dyes. Each plate is scanned twice to get Cy3 emission (green) and Cy5 (red) emission.

Step 5: Data analysis and interpretation

The fluorescent image of the plate is captured and stored as raw data in 16-bit TIFF file. The raw data is normalized to remove experimental errors. The background fluorescence of Cy3 and Cy5 must be

Published by Woodhead Publishing Limited

subtracted. Background subtraction can be done either for individual spots or the whole plate. Spot showing weak intensities are eliminated from the analysis. From the spot intensity (ratio of Cy3 to Cy5 fluorescence), genes that are up or down regulated are identified. From the uniqueness of spot fluorescence (Cy3 or Cy5), tissue-specific gene expression is identified.

Microarray data analysis tools

1. TreeView
2. Expressionist
3. GeneTraffic
4. Spotfire
5. GeneSpring

Review questions and answers

1. *Serial analysis is a high throughput transcriptomics technique. Substantiate the statement.*

Expression of many genes can be studied from a single SAGE experiment.

2. *List the limitations of SAGE.*

- It requires completely sequenced genomes for the study of tags comparison and gene identification.
- It requires high throughput sequencing facility.
- Low abundant cDNA tags are often missed.
- Wrong identification of genes due to sequencing error or false matching of the tags with gene sequences.
- It requires a relatively high amount of initial mRNA (2.5–5.0 µg of poly (A)⁺ RNA).

3. *SAGE experiments require a high quality and quantity of starting mRNA. This is not possible sometimes. Suggest some other alternatives to the SAGE method.*

Modification of SAGE protocol requiring less quantity of initial mRNA from the sample is called microSAGE. It requires only 500-fold less mRNA than the usual SAGE experiment.

4. *What are SAGE tags? How are they useful in identifying a transcript?*

9–12 nucleotides attached to the 3' end of the cDNA are called a tag. It has been shown that a small sequence of the cDNA known as the 'SAGE tag' is sufficient to identify the gene. The gene corresponding to the SAGE tag is identified by searching against the cDNA or EST databases.

5. *What is an expression profile?*

It is a catalogue of all genes expressed in a cell or tissue.

6. *Can you correlate the transcriptome and proteome of a cell?*

We cannot make a direct correlation between the transcriptome and the proteome as post-trancriptional and postranslational modifications change the relationship.

Recommended reading

Albelda, S.M. and Sheppard, D. (2000) 'Functional genomics and expression profiling: be there or be square', *American Journal of Respiratory Cell and Molecular Biology*, 23: 265–69.

Bauer, D. et al. (1994) 'Detection and differential display of expressed genes by DDRT-PCR', *Genome Research*, S97–S108.

Bilban, M. et al. (2002) 'Normalizing DNA microarray data', *Current Issues in Molecular Biology*, 4: 57–64.

Datson, N.A. et al. (2010) 'MicroSAGE: a modified procedure for serial analysis of gene expression in limited amounts of tissue', *Nucleic Acid Research*, 27:1300–7.

Hubank, M. and Schatz, D.G. (1994) 'Identifying differences in mRNA expression by representational difference analysis of cDNA', *Nucleic Acid Research*, 22: 5640–8.

Moody, D.E. (2001) 'Genomics techniques: an overview of methods for the study of gene expression', *Journal of Animal Science*, 79(Suppl.): E128–E135.

Yamamoto, M., Wakatsuki, T., Hada, A. and Ryo, A. (2001) 'Use of serial analysis gene expression (SAGE) technology', *Journal of Immunological Methods*, 25: 45–66.

Web addresses

http://www.ebi.ac.uk/arrayexpress/.
http://www.medicalgenomics.org/databases/blotbase/news.
http://www.ncbi.nlm.nih.gov/geo/.
http://www.ncbi.nlm.nih.gov/SAGE/.
http://www.uk.sagepub.com/error404.nav.

Published by Woodhead Publishing Limited

10

Introduction to proteomics

To really understand biological processes, we need to understand how proteins function in and around cells since they are the functioning units.

Hanno Steen, Director of the Proteomics Center
at the Children's Hospital, Boston

Abstract: The final product of gene expression is protein. Proteins are the functional entities of cells. The total number of proteins expressed in a cell is called a proteome. This chapter describes the importance of proteomes and proteomics techniques. Different protein isolation methods are analyzed. Important proteomics techniques are discussed. Different branches of proteomics such as structural proteomics, functional proteomics are also introduced.

Key words: 2D-PAGE, mass spectrometry, proteome, proteomics, SDS-PAGE.

Key concepts

- Proteomics is the large-scale study of all proteins expressed in a cell or tissue.
- Major techniques for protein separation are SDS-PAGE, 2D-PAGE and liquid chromatographic methods.
- SDS-PAGE is the most widely used technique for large-scale protein separation, but it separates only a limited number of proteins in a single gel.
- The 2D-PAGE technique is capable of separating thousands of proteins in a single gel.
- Mass spectrometry analysis combined with 2D-PAGE has become a routine high throughput protein separation and identification method.
- Both 2D-PAGE and mass spectrometry are used for quantitative proteomics study.

10.1 Introduction

Proteins earned their primary responsibility in cellular function even before DNA. Proteins are the most complex of all biological molecules in terms of their structures and functions. There is no place in a cell which functions without proteins. The structure of proteins ranges from primary to quaternary structures. Proteins are synthesized according to the information present in the DNA fragment (gene) but for the synthesis of DNA (replication), proteins (DNA polymerase) are required. Proteins rarely work alone in the cells. The cross-talk between proteins and other macromolecules like DNA, carbohydrate, lipids and other proteins is exemplary.

Proteins are the most dynamic entities and the real executers of cellular functions. Although DNA is the real storehouse of genetic information, the information has to be transcribed and translated into proteins. Proteins follow a universal pathway from synthesis to degradation. Protein synthesis is termed **translation**; proteins are modified to take up different tasks. Nascent polypeptides undergo a **protein folding** pattern to acquire secondary and tertiary structures which are required to form a biologically active protein. Fully matured and modified proteins are ready for targeting to different places in a cell to perform their functions.

10.2 Traditional route of protein study

The traditional method of studying proteins involved isolating a particular protein by solvent precipitations or the salting out method. The precipitated proteins were analyzed for their quality and relative molecular weight using SDS-PAGE analysis and the proteins were characterized for amino acid composition and amino acid sequencing by **Edman degradation**. This kind of study is known as protein chemistry. It is essentially dedicated to studying one protein only.

10.2.1 Proteome and proteomics

The total number of proteins expressed in a cell at a particular time is known as a proteome. The status of the cell can be assessed by its proteome, for example, the proteome of a healthy cell is different from that of a diseased cell. After **genomics**, proteomics has become the fundamental aspect of studying biological system. Proteomics is the study of an entire complement of proteins expressed in a particular cell at a particular time. There are different branches of proteomics, such as structural proteomics, cellular proteomics, functional proteomics, phosphoproteomics, or glycoproteomics. Structural proteomics involves the study of high throughput three-dimensional structure determinations of proteins. Quantitative proteomics involves the relative quantification of proteins in two different samples under study. Actually, the number of proteins expressed in a cell exceeds the total number of genes present in an organism. For example, in a human being, the estimated number of protein coding genes ranges from 25,000 to 30,000 but the number of proteins that are expressed in human cells is ten times more.

The terms proteome and proteomics were first described by Marc Wilkins and his colleagues in 1995 by combining the words *prote*in and gen*ome* and so the study of proteome is known as proteomics.

Protein separation and identification techniques are central aspects of proteomics research. Proteomics techniques are different from traditional protein chemistry in the following ways:

1. Use of high throughput separation techniques in which a large number of proteins are separated in a single platform.
2. Sensitive detection methods have the ability to detect low abundant proteins.
3. Analysis of the result with protein and nucleic acid databases helps us to identify the proteins faster.

10.3 Protein isolation methods

Cell lysis is the primary step in fractionation of proteins. The best possible method has to be identified for the maximum yield and purity. The extracted protein sample is stored at −80°C for further use. For proteomics study, proteins extracted from fresh samples is preferred. To avoid protein degradation, a cocktail of protease inhibitors is added, and also the sample should be extracted at a low temperature preferably using liquid nitrogen. The protein of interest should be separated from other contaminant proteins for the characterization, three-dimensional structure and interactions. Different steps are followed to separate one particular protein from the mixture of the others. A minimum number of purification steps not only reduces the cost of the final product but also increases the yield. The protein of interest from microbial source may be intracellular, membrane-bound or extracellular. For example, disintegration techniques are required only for membrane-bound or intracellular proteins but not for extra-cellular proteins. Purification strategies have to be modified depending on the source. The choice of the purification step is also based on the size and the physical and chemical properties of the protein.

10.3.1 Extraction methods

If the desired protein is extra-cellular, filtration or centrifugation removes the microbial biomass. Similarly, the enzyme is intra-cellular or membrane-bound; the collected biomass should be broken to release the enzyme. Filtration and centrifugation are the most commonly used methods to separate the extra-cellular proteins from the fermentation broth. If the protein is intra-cellular, the cells should be collected from the broth for further processing.

Tissue homogenization

This is done using a tissue grinder to homogenize and de-aggregate animal or plant tissues mechanically to obtain cells. This is done by applying shear forces to the materials which are preferably placed in a buffer. Tissue homogenization can be scaled up easily and can handle large volumes. Rotating blades in a mixie/Waring blender break up the tissues into cells.

Grinding with acid-washed sand or glass beads

Microbial biomass or plant/animal cells (at 4°C) are placed in a mortar containing acid-washed sand or glass beads. The cells in a mortar are ground to a fine paste using a pestle and a sufficient amount of buffer is added to dissolve the paste. The contents are then centrifuged at 10,000 rpm for 15 minutes to obtain a cell-free extract. This is an inexpensive method and easy to perform at laboratory level.

Freezing and thawing method

This method is suitable for the lyses of animal cells and Gram-negative bacterial cells. The sample to be disrupted is placed in a buffer which is in turn placed in the freezer compartment of a refrigerator (at –4°C or –20°C) or a liquid nitrogen container. The frozen sample is taken out of the freezer to thaw at 37°C. The formation of ice crystals during freezing pierces the cells and the cells also shrink during thawing. The above procedure is

Published by Woodhead Publishing Limited

repeated quite a few times to release more of the intracellular contents. This method is not effective for Gram-positive bacterial cells and plant cells.

Ultrasonication

This is carried out using an ultrasonicator with a microtip. The temperature of the sample should be at 0°C, preferably in ice containers. The sample is ultrasonicated continuously for no more than 20 seconds or in pulses and then the sample is kept at 4°C for 30 seconds for cooling. This procedure is repeated six to seven times for complete disruption. The duration of the ultrasonication depends on the nature of the cells. The ultrasonic waves create microscopic vapour bubbles which, when they collide on cells, cause their disruption. It is difficult to scale up the ultrasonication and it is expensive.

Centrifugation

Centrifugation exploits the density difference between the protein molecules to be separated and the fermented broth. Centrifuges are frequently used to separate solids from liquids. Centrifuges are ideal candidates for the removal of bacterial biomass in the fermentation broth or to sediment the protein precipitates that are obtained during ammonium sulfate fractional precipitation or solvent (e.g. chilled acetone). If the enzyme is extra-cellular in nature, the microbial biomass in the fermented broth should be removed by centrifugation. Refrigerated centrifuges (high speed centrifuges and ultracentrifuges) are preferred for the purification of enzymes.

10.3.2 Protein separation techniques

SDS-PAGE

Sodium Dodecyl Sulfate PolyAcrylamide Gel Electrophoresis (**SDS-PAGE**) is an electrophoretic technique widely used in biotechnology, biochemistry, molecular biology, forensic science and other life science laboratories. In SDS-PAGE, proteins are separated in a palyacrylamide gel based on their molecular weight. Proteins are amphoteric molecules, i.e. they have both positive and negative charges. To make them move in a single direction, a uniform negative charge is created on them. When the proteins are mixed with SDS, they acquire a net negative charge.

SDS is a detergent having a negative charge, therefore it is an anionic detergent. SDS denatures the native proteins by disturbing the non-covalent forces. The non-covalent forces include hydrogen-bonding, hydrophobic and ionic interactions which are responsible for the three-dimensional structure of a native protein. SDS also gives a uniform net negative charge to the protein molecules. The denatured linear protein molecules are loaded onto the polyacrylamide gel (PAGE) which is made by polymerizing the acrylamide monomers. PAGE is prepared using acrylamide, bisacrylamide, TEMED, ammonium persulfate and Tris-HCl buffer. PAGE has two phases: a stacking gel and a separating gel. Under an applied electric field, the stacking gel concentrates the SDS-loaded linear protein molecules while the separating gel separates the proteins on the basis of molecular weight. After the run, PAGE gel is placed in a Coomassive Brilliant Blue R250 dye solution for staining for a few hours and is de-stained to visualize the separated protein molecules as bands.

2D-PAGE

Two-dimensional gel electrophoresis or **2D-PAGE** is the primary technique for proteomics work. It separates the complex mixture of samples using two different properties of the proteins. In the first dimension, proteins are separated by the pI value and in the second dimension by the relative molecular weight. Although it was described way back in 1975 by O'Farrell, its applicability and adoptability were enhanced because of the introduction of immobilized pH gradient strips, as they gave good reproducible results and handling became easy. Initially proteins were visualized by using ^{32}P or ^{35}S labelling. Now this has been replaced by more sensitive techniques such as SYBRO Ruby. Advances were made in different stages of the 2D-PAGE technique which are capable of separating up to 10,000 proteins in a single gel. A 2D-PAGE gel image is captured and image analysis is done to find the number of proteins expressed in a particular tissue.

Mass spectrometry

Mass spectrometric (MS) analysis of proteins has become an important tool in proteomics study. The basic principles and different types of mass spectrometers are explained in Chapter 12. In short, MS analysis involves separation of gas phase-charged molecules based on their mass to charge ratio. Proteins are highly sensitive biomolecules. They require mild conditions for analysis. Proteins carry charge but their net charge varies with the type of

Published by Woodhead Publishing Limited

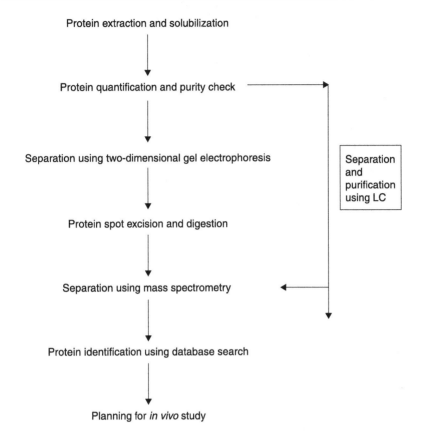

Protein extraction and solubilization

Protein quantification and purity check

Separation using two-dimensional gel electrophoresis

Separation
and
purification
using LC

Protein spot excision and digestion

Separation using mass spectrometry

Protein identification using database search

Planning for *in vivo* study

Figure 10.1   Flow chart for any proteomics study.

amino acid sequence and the buffer in which they are dissolved. The most commonly used MS types for protein analysis are **MALDI** and ESI ion sources. MS analysis of protein is helpful in accurate molecular weight determination, protein identification through peptide mass finger printing, and *de novo* **protein sequencing** using tandem analysis. It is also possible to map the post-translation site modifications like phosphorylation and glycosylation. Figure 10.1 shows the flow chart for any proteomics study.

10.4 Branches of proteomics

1. *Structural proteomics*: deals with high throughput structure prediction methods for all proteins.
2. *Functional proteomics*: deals with high throughput function prediction methods for all proteins.

3. *Quantitative proteomics*: deals with high throughput quantification of methods for all proteins.
4. *Glycoproteomics*: deals with high throughput glycoprotein analysis methods.
5. *Phosphoproteomics*: deals with high throughput phosphoprotein analysis methods.
6. *Expression proteomics*: deals with the number of proteins that are expressed in different cell types in an organism.
7. *Clinical proteomics*: deals with analysis of different proteins in biological samples such as blood, urine, etc., to identify biomarkers for diagnosis and drug targets.

10.4.1 Quantitative proteomics

This involves the quantitative measure of proteins expressed in a cell in a particular condition. It is basically a relative quantification rather than an absolute quantification. It is important to understand the protein functions and their abundance. Present methods used for quantitative proteomics are 2D-PAGE or 2D-PAGE followed by MS analysis, protein array-based, and MS signal intensity-based quantification. The combination of 2D-PAGE followed by MS is the widely followed method for quantitative proteomics study. Since quantification is a relative expression between healthy vs diseased, treated vs control, protein samples from the two sets of samples have to be separated in 2D-PAGE and by comparing the spot position and intensity, differentially expressed proteins in these gels are identified. But results from 2D-PAGE alone cannot be taken as final, as the reproducibility of this technique is low. To increase the sensitivity of the protein detection, protein samples can be labelled with two different fluorescent dyes and the image can be analyzed. 2D-PAGE followed by mass spectrometric analysis will give a better result than 2D-PAGE analysis alone. Proteins can be labelled by two methods: (1) *in vivo* labelling; and (2) *in vitro* labelling. *In vivo* labelling involves metabolic incorporation of stable isotopes in the proteins. The most commonly used technique is incorporating the ^{15}N or ^{14}N nitrogen in the form of ammonium sulfate in the culture medium. A relative quantity of the protein can be calculated from the mass shift in the mass spectrum. Another method for metabolic labelling is Stable Isotope Labelling with Amino acids in Culture (SILAC). Isotopically labelled amino acids like ^{2}H-Leucine, ^{13}C Lysine are used. This method has one advantage as labelling is done even before protein synthesis and this reduces the handling variations. However, this has one major limitation: It cannot be used for samples which are not grown in the *in vitro* medium.

10.4.2 Protein–protein interaction

Proteins rarely work alone; many of the cellular processes are carried out by groups of proteins. Replication, transcription and translation are the fundamental cellular processes which involve many proteins working together or sequentially. Although protein–protein interaction was known, study on this was limited. The branch of proteomics which deals with protein–protein interaction is called interactome. Depending upon the type of protein association, there are two types; homodimer, if two sub-units of the same polypeptides combine and form an active complex, or heterodimer, a protein–protein interaction state in which two different polypeptides are involved. Protein–protein interaction is necessary for the efficient and economic functioning of the system. Protein–protein interaction can be studied using many techniques such as protein affinity chromatography, co-immuno precipitation, yeast two hybrid and phage display.

10.5 Characteristics of proteomics

Any protein will have the following characteristics, such as amino acid sequence, isoelectric point, molecular weight, secondary structure and its associated functions, post-translational modification, etc. Table 10.1 presents a comparison between genomics and proteomics.

Applications of proteomics

1. Molecular weight determination.
2. Isoelectric point determination.
3. Amino acid sequence.
4. Quantification of proteins.
5. Post-translational modification.
6. Peptide mapping.
7. Comparative proteomics.
8. Structure determination.
9. Biomarker identification.
10. Drug–target identification.

Published by Woodhead Publishing Limited

Table 10.1   Genomics and proteomics: a comparative analysis

Sl. No.	Genomics	Proteomics
1	Entire DNA present in a cell	Entire complement of proteins expressed in a cell
2	It is the storehouse of genetic information	They are the cellular executers of cellular functions
3	It more or less does not change throughout the life of an organism; there is not much change in the amount of genome	Changes according to the environment, it is the most dynamic macromolecule in living cells
4	It is not complex, usually one set of genome will be there for an organism	It is more complex; a single organism will have more than ten different proteomes, like phosphoproteome, glycoproteome, etc.
5	Represented as string of nucleotides, A, T, G, C	Represented as a string of amino acids of 20 different types
6	The nature of the gene sequence is not altered by the cellular environments	It is highly altered by the cellular environments
7	Genomics study gives indirect information about cell functions	Proteomics study gives more direct information about the cell functions
8	It is easy to study and direct methods are established	It is difficult to study as different methods are required to study different proteomics

Review questions and answers

1.   *What is the difference between protein chemistry and proteomics?*

Proteomics techniques are different from traditional protein chemistry in the following ways:

- Use of high throughput separation techniques so that a large number of proteins is separated in a single platform.
- Sensitive detection methods have the ability to detect low abundant proteins.
- Analysis of results with protein and nucleic acid databases to identify the proteins faster.

2. *List some proteomics sites.*

    http://www.expasy.ch/ch2d/.
    http://prospector.ucsf.edu/.
    http://thompson.mbt.washington.edu/sequest/.
    http://protein.toulouse.inra.fr/prodom.html.

3. *What are the different branches of proteomics?*

- Structural proteomics.
- Phosphoproteomics.
- Glycoproteomics.
- Quantitative proteomics.
- Functional proteomics.

Recommended reading

Bharati, M.D., Shweta, P.V., Priya, C.B., Namita, K.A., Suvarna, D.G., Amol, B., Pallavi, K. and Gomase, V.S. (2009) 'Proteomics: emerging analytical techniques', *International Journal of Genetics*, 1: 17–24.

Crawford, M.E., Cusick, M.E. and Garrels, J.I. (2000) 'Databases and knowledge resources for proteomics research', *Proteomics: A Trends Guide*, 4: 17–21.

Crista, I.M., Gaskell, S.J. and Whetton, A.D. (2004) 'Proteomics techniques and their application to hematology', *Blood*, 103: 3624–34.

Gevaert, K. and Vandekerckhove, P. (2000) 'Protein identification methods in proteomics', *Electrophoresis*, 21: 1145–54.

Matt, P., Fu, Z., Fu, Q. and Van Eyk, J.E. (2008) 'Biomarker discovery: proteome fraction and separation in biological samples', *Physiological Genomics*, 33: 12–17.

Patterson, S.D. and Aebersold, R.H. (2003) 'Proteomics: the first decade and beyond', *Nature Genetics Supplements*, 33: 311–21.

Wilkins, M.R. et al. (1995) 'Progress with proteome projects: why all proteins expressed by a genome should be identified and how to do it', *Biotechnology and Genetic Engineering Reviews*, 13: 19–50.

Websites

See Table 10.2 for a list of URLs for proteomics study.

Published by Woodhead Publishing Limited

Table 10.2 List of URLs for proteomics study

Sl. No.	Name of the site	URL	Dedicated work
1	NCBI/BLAST tools	http://ncbi.nlm.nih.gov/BLAST/	Sequence data base
2	SWISS-2DPAGE	http://www.expasy.ch/ch2d/	Protein identification
3	Peptident	http://www.expasy.ch/tools/peptinent.html	Protein identification
4	Protein Prospector	http://prospector.ucsf.edu/	Protein identification
5	SEQUEST	http://thompson.mbt.washington.edu/sequest/	Protein identification
6	ProDOM	http://protein.toulouse.inra.fr/prodom.html	Domain identification
7	2D PAGE	http://web.mpiib-berlin.mpg.de/cgi-bin/pdbs/2d-page/extern/	2D-PAGE images
8	SWISS-MODEL	http://scop.mrc-lmb.cam.ac.uk/scop/	3D structure analysis
9	Protein Database (PDB)	http://www.rcsb.org/pdp/	3D structure analysis

11

Two-dimensional gel electrophoresis of proteins

Abstract: Proteomics is a new field which developed after the genome sequencing projects. The proteomics techniques developed are centred on a core technique called two-dimensional gel electrophoresis or 2D-PAGE. As the name indicates, the proteins are separated in two directions which are orthogonal to each other, based on two properties of the proteins such as the isoelectric point and molecular weight. This technique is capable of separating as many as 10,000 proteins in a single gel. In this chapter, the brief history and different steps involved in 2D-PAGE separation are discussed in detail.

Key words: 2D-PAGE, ampholytes, equilibration, IEF, image analysis, IPG strips, isoelectric point, SDS-PAGE.

Key concepts

- **2D-PAGE** is a high throughput protein separation technique which is capable of separating up to 10,000 proteins in a single gel.
- The separation of proteins is based on two different properties of the proteins, that is, the isoelectric point and the molecular weight.
- Before **isoelectric focusing,** the sample protein to be separated must be completely solubilized using denaturing agents like urea/thiourea, non-ionic detergents and reducing agents like DTT.
- Isoelectric focusing is the first-dimensional separation based on the isoelectric point (pI)
- Second-dimensional separation is done similar to the SDS-PAGE of the Laemmli protocol.
- 2D-PAGE separation of protein shows spots and each spot is characterized with specific pI and relative molecular weight values.

Published by Woodhead Publishing Limited

11.1 Introduction

The entire complement of proteins expressed in a cell is known as the proteome. A proteome is further complicated by post-translational modifications. Conventional single dimension separation of proteins resolved only a limited number of proteins. Only very few proteins can be separated using **SDS-PAGE**. But even the smallest viral genome codes have more than five hundred proteins. Therefore, it was important to have a technique which could separate the number of proteins and two-dimensional gel electrophoresis is one such technique. A proteome is highly dynamic in the sense that it responds to external and internal stimulus. Proteins execute their functions differentially depending upon their folding state, phosphorylation state, glycosylation state, etc. Alternative splicing is the major phenomenon which determines the different types of proteins coded from the same gene. Proteomics involves qualitative and quantitative analysis of proteins.

11.2 Principles of 2D-PAGE

A complex mixture of proteins is separated based on two parameters such as the isoelectric point and the molecular weight in two different directions. Proteins are separated in the first dimension based on their isoelectric point. Proteins separated in the first dimension are placed on the separating gel of the SDS-PAGE (Figure 11.1). In the second dimension,

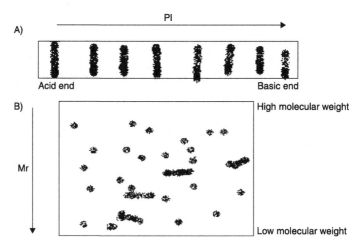

Figure 11.1 Principles of 2D-PAGE separation of proteins. A) First dimension: isoelectric focusing based on isoelectric point. B) Second dimension: SDS-PAGE separation based on molecular weight.

proteins are separated according to their molecular weight. Since proteins are separated based on two parameters, they are separated all over the gel.

Proteins are amphoteric biomolecules, i.e. they bear both positive and negative charges. Depending upon their amino acid sequence, they carry a positive charge or negative charge. The net charge of a protein is affected by the surrounding solution or buffer. Based on the pH of the solution, proteins undergo either protonation or deprotonation and have either net positive or a net negative charge. Depending on their net charge, proteins move towards either the cathode or the anode. At a particular pH, the positive and negative charge become equal and the proteins cannot move in either direction, therefore, they are said to be focused. This technique is called isoelectric focusing. The **isoelectric point** (pI) is defined as the pH at which the protein has no net charge. The charge on the protein under a particular pH is calculated using the Henderson–Hasselbach equation.

11.3 2D-PAGE apparatus

The 2D-PAGE apparatus consists of two assembly units: (1) the isoelectric focusing unit (IEF Unit); and (2) the second dimensional unit.

11.3.1 The IEF unit

In this unit, proteins are separated based on their isoelectric point. In the early days, capillary tubes were used to prepare tube gels for the first-dimensional run. Later, IPG strips were developed. Commercially available IEF units are suitable for IPG strips. The unit is connected to circulating waterbath which is maintained at 16°C to prevent the heat generation due to the high voltage used. It has a metal platform on which a glass tray is fixed to hold the IPG strips.

11.3.2 The second dimensional unit

This is similar to the Laemmli vertical apparatus. The vertical glass plates are assembled to prepare the separating gel. Usually this unit is connected to circulating waterbath maintained at a low temperature.

Published by Woodhead Publishing Limited

11.4 Sample preparation

Cell lysis and protein solubilization are the key steps in proteome sample preparation. Complete recovery of proteins is essential for successful proteome study. Care should be taken to include the low abundant proteins also. It also involves the use of certain reagents in the upstream and the downstream steps. The sample to be used for the protein separation should be representative, i.e. it should not have any heterogeneous cell types. This is achieved by using laser capture micro-dissection techniques. Although sample preparation for proteomics involves normal disruption and fractionation techniques, additional care should be taken.

The samples after collection from the sources should be stored at −80°C till use. In order to avoid protein degradation, samples should be ground with liquid nitrogen into a fine powder. Utmost care should be taken while extracting the proteins from the sample; the quality of the protein affects the IEF of the proteins in the first dimension which in turn affects the second-dimensional separation and spot identification. The powdered sample is immediately transferred to the sterile tube containing a sample extraction buffer which consists of 8 M urea, 2 M thiourea, 2 per cent CHAPS, 50 mM DTT and 0.2 per cent ampholyte. The sample is re-suspended completely and centrifuged at 15,000 rpm for 30 min. to get a clear supernatant containing almost all the proteins of the sample. The clear supernatant is transferred to a new tube and stored for further analysis.

Salts and buffer components should be kept at a minimum concentration. Higher salt concentration affects the separation of protein during IEF. The sample should be completely solubilized before going for IEF. Otherwise it leads to horizontal streaking in the second-dimensional SDS-PAGE gel. If the salt concentration is high (that is, more than 10m M), the sample should be dialysed. Precipitation with acetone or trichloroacetic acid can also be performed to remove the salts and this method also removes unwanted small molecules, lipids, nucleotides. This desalting procedure works well, although some amount of the proteins will be lost in the procedure. Proteins will be solubilized in the presence of detergents, but one cannot use detergents like SDS at more than 0.25 per cent. Instead, non-ionic detergents like CHAPS can be used. The amount of sample to be loaded depends on the size of the IPG strip. An optimum sample preparation is aimed at solubilizing all the proteins and keeping them in a solubilized state.

Protein extraction methods also depend on the aim or goals of the experiment being planned. Rehydration is done to prepare the immobilized strips for sample loading. This consists of urea and thiourea which act as choatropic agents. Detergents are important to prevent the protein aggregation and thereby increase the solubilization. The addition of Zwitterionic detergents like CHAPS, ASB14, increases the solubility and the number of spots that can be detected.

Sample solubilization buffer	
Urea	8 M
Thiourea	2 M
CHAPS	2%
DTT	50 mM
Ampholyte	0.2%

11.5 First-dimensional separation by isoelectric focusing

11.5.1 Preparation of pH gradient for IEF

The pH gradient for IEF can be prepared in two ways. Originally, the pH gradient was made in the tube gels by mixing the synthetic carrier ampholytes along with an acrylamide gel mixture. These synthetic carrier ampholytes are small molecules with different carboxylic and amino groups and are available as a mixture of individual units covering a specific range of pH. A SCA mixture is mixed with a polyacrylamide stock solution and the glass capillary tube filled, which is then allowed to solidify. During the pre-run, the SCA will migrate according to their charge and reach their respective pI value. The SCA with the lowest pI will be nearer to the cathodes and the SCA with the highest pI will be nearer to the anode. The rest of the SCA units will form a continuous pH gradient between the anodic and cathodic ends. The unique property of these SCAs is that they show very high buffering capacity at their pI value. But many problems are encountered with SCA. Since SCA molecules are not firmly attached to the gel, their migration is not controlled. Therefore, this results in poor reproducibility of the migration pattern between runs. Another problem with SCA is that the pH gradient is not correctly maintained at the cathodic end, therefore, protein focusing at the cathodic end is not clear and this is known as cathodic drift. Separation of very basic proteins is a major problem with SCA. The tube gels are very fragile and difficult to handle.

Published by Woodhead Publishing Limited

In 1985, immobilized pH gradient (IPG) strips were introduced in place of SCA performed with tube gels. IPG strips are prepared by mixing a special kind of acrylamide in which SCA is covalently attached. The gel is prepared in acrylamido buffers which provide the pH gradient and are cut into small strips of 3 mm width and varying lengths which are known as IPG strips. There are two kinds of IPG strips: linear (L) pH gradient strips and non-linear (NL) pH gradient strips.

IEF with IPG strips provides reproducible results as the pH gradient is fixed and also it works without any cathodic drift. Since these strips are prepared on a plastic sheet, handling is easy (no breakage). These strips have higher sample loading capacity, because of in-gel rehydration, are very acidic and basic proteins can also be separated. Pre-cast IPG strips are commercially available in different lengths (7 cm, 11 cm, 13 cm and 24 cm) and pH ranges. Longer strips and a higher pH range are recommended for better separation of total proteins for isoelectric point determination.

11.5.2 Sample loading

The first-dimensional run is performed to separate the proteins based on their isoelectric points. In the case of tube gels, a sample application is done at the top of the tube gel and the IEF is carried out. In the case of the IPG strips, rehydration is done before proceeding to IEF as they are stored in a dried condition. The usual composition of rehydration solution contains urea/thiourea, detergent (CHAPS), reducing agent (DTT) and ampholyte solution. Rehydration is performed in a re-swelling tray in which the IPG strips are layered in a rehydration buffer, and silicon oil is used to prevent evaporation. Rehydration is done at room temperature for 45 minutes. to one hour. Sample loading is done in two ways: (1) cup-loading; and (2) in-gel rehydration. In-gel rehydration is preferred to cup-loading because this method allows the loading of more quantity of sample. Even when diluted, the samples can be used for IEF.

11.5.3 Isoelectric focusing

After rehydration and sample application, the IPG strips are carefully taken out of the rehydration buffer and the oil is drained off. The IPG strips are aligned on the IEF platform and electrodes are placed on the strips. Usually IEF is done at high voltage. The entire unit is connected to circulating waterbath which is maintained at 16°C. The extent of electrophoresis can be traced with the help

Published by Woodhead Publishing Limited

of a migrating dye (bromophenol blue). After IEF, either the strips are used for second-dimensional SDS-PAGE or stored at –70°C for weeks.

11.6 Equilibration

After IEF, the IPG strips are saturated with SDS to give a net negative charge to the proteins, which is required for second-dimensional separation. This is done immediately prior to the second-dimensional separation. The equilibration buffer contains urea, glycerol, reductant, SDS and bromophenol blue. The equilibration solution contains 50 mM Tris-Cl which is required for second-dimensional electrophoresis. Glycerol helps in transferring proteins from the IPG strip to the second-dimensional gel. An additional equilibration step is also done with iodocetamide which alkylates the thiol groups of proteins. Protein oxidation is a major problem during electrophoresis as this results in streak formation in the final image.

Equilibration buffer

Urea
Glycerol
DTT
SDS
Bromophenol blue

11.7 Second-dimensional separation by SDS-PAGE

After equilibration, the IPG strips are ready for second-dimensional separation; this is similar to the Laemmli protocol for SDS-PAGE. The polyacrylamide gel is cast with the same composition and the size of the gel is decided according to the length of the IPG strip. After polymerization of the separating gel, the equilibrated IPG strip is aligned on the surface of the gel and firmly fixed with the help of agarose gel. A protein molecular weight maker can be loaded at the corner of the separating gel. The electrophoresis is carried out till the tracking dye reaches the bottom. The separating gel is carefully removed and washed with distilled water. The unpolymerized gel mixtures should be removed before proceeding to staining. Different steps involved in 2D-PAGE are shown in Figure 11.2.

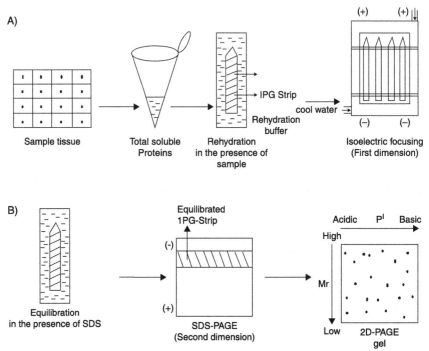

Figure 11.2 Steps involved in two-dimensional gel electrophoresis. A) Total protein extracted from the sample is applied on the IPG strip and subjected to first-dimensional IEF. B) The IPG strips are equilibrated in the presence of SDS to make all the protein negatively charged. Then the strip is aligned on the separating gel and second-dimensional SDS-PAGE is performed. Proteins are detected as spots all over the gel surface after staining.

11.8 Detection of proteins on 2D-PAGE gels

Various staining techniques are used to locate the protein spots on the 2D-PAGE gel. Some of the widely used staining dyes are Coomassie Brilliant Blue (CBB), silver stains, SYBRO Ruby. The selection of staining technique depends on many criteria such as sensitivity, compatibility with MS analysis and the dynamic range of the dye. Although silver staining is sensitive up to the lower level of 10ng per spot, it is not widely used because it is incompatible with MS analysis. CBB is cheap and best suited for MS analysis of proteins for peptide mass fingerprinting. SYBRO Ruby is a fluorescent dye and sensitive enough to detect one ng per spot. Sometimes double staining can be done in a sequential manner to detect the spots unstained with one dye. Double staining is done in the sequence of SYBRO Ruby followed by CBB to detect low abundant proteins which are not detected by CBB.

Staining techniques for 2D-PAGE gel

1. Coomassie brilliant blue
2. Silver nitrate staining
3. SYBRO Ruby

11.9 Image analysis

The total number of spots which appears on a 2D-PAGE gel is around 2000–4000 depending upon the sample type, extent of separation and staining method. If the proteins are well separated, each spot represents one polypeptide. The shape of spots varies and they may be circular, oval, elliptical or irregular. It is very difficult to analyze the 2D-PAGE gel manually. The gel image is captured using a CCD camera, fluorescence detectors, and scanning densitometers, depending upon the staining technique employed. The analogue image is converted to a digital image. Therefore, a 2D-PAGE gel image analysis is done using a computer software program. Spot identification is complicated by background noise and streaks in the gel. Quantification of the proteins can be done by measuring the spot intensity. This depends on the spot density which in turn depends on the type of staining method used. The intensity percentage of each spot can be measured as relative intensity. The steps in image analysis follow image cleaning. Image cleaning involves removal of background colour and vertical and horizontal streaks. In the next step, the software counts and reports the number of spots present in the gel. It is also possible to match the gel images of related samples and identify the differentially expressed proteins. Figure 11.3 shows the steps in image analysis.

Software packages for 2D-PAGE image analysis

1. PDQuest (Bio-Rad, Hercules, CA, USA)
2. Melanie 3 (GeneBio, Geneva, Switzerland)
3. Delta2D (DECODON, Greifswald, Germany)
4. ImageMaster 2D (Amersham BioSciences)
5. Bioimage 2D investigator (Genomic Solutions)

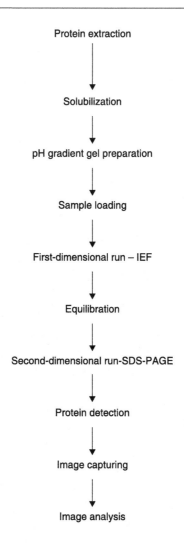

Protein extraction

↓

Solubilization

↓

pH gradient gel preparation

↓

Sample loading

↓

First-dimensional run – IEF

↓

Equilibration

↓

Second-dimensional run-SDS-PAGE

↓

Protein detection

↓

Image capturing

↓

Image analysis

Figure 11.3 Steps in image analysis.

11.10 Application of 2D-PAGE in proteomics

2D-PAGE is capable of separating thousands of proteins in a single gel. Therefore, it is a protein expression profiling technique. The total number of proteins expressed in a cell type can be catalogued. This catalogue can also be used to find differentially expressed proteins between related samples, for example, healthy vs diseases, treated vs control. From the

2D-PAGE image, it is possible to find the protein expression status of tissue. Therefore, it can be used to find out proteins that are expressed only under particular condition. Proteins separated on 2D-PAGE are nearly pure, therefore, they can be excised and used in the **Edman degradation** method of protein sequencing and for mass spectrometric analysis. 2D-PAGE images can also be used to detect certain post-translational modifications like glycosylation, phosphorylation, etc.

Review questions and answers

1. *What is the principle of 2D-PAGE separation?*

This is a separation technique in which two different properties of the proteins are exploited to separate them. Proteins are separated in the first dimension based on their isoelectric point, then subsequently separated by molecular weight in a single gel.

2. *What are synthetic carrier ampholytes and how do they help to create pH gradient?*

Synthetic carrier ampholytes are small molecules with different carboxylic and amino groups. These SCAs are available as a mixture of individual units covering a specific range of pH. The SCA mixture is mixed with a polymerization mix, poured into a glass capillary tube and allowed to solidify. During the pre-run, the SCA units will migrate according to their charge and reach their respective pI value, and create a pH gradient.

3. *IPG strips are better than SCA. Why?*

Because the IPG strips form a stable pH gradient.

4. *2D-PAGE separation of proteins requires elaborate sample preparation method. Why?*

For IEF and second dimension electrophoresis, the sample should be free of salts and ions, and they must be removed by dialysis.

5. *Silver staining sensitive detection method is used for proteins, but is commonly not preferred. Why?*

Silver-stained proteins are not compatible with mass spectrometry analysis.

6. *2D-PAGE is an excellent technique capable of separating thousands of proteins. However, its use is limited. Why?*

Although 2D-PAGE separates thousands of proteins in single gel, the reproducibility of a result is the major problem.

Recommended reading

Chrishtof, R. and Srinivasan, M. (2006) 'The application of proteomics to plant biology: a review', *Canadian Journal of Botany*, 84: 883–92.

Czegledi, L., Gulyas, G., Radacsi, A., Kusza, S., Bekefi, J., Beri, B. and Javor, A. (2010) 'Sample preparation and staining methods for two-dimensional polyacrylamide gel electrophoresis of proteins from animal tissues', *Animal Science and Biotechnologies*, 43: 267–70.

Dowsey, A.D., Dunn, M.J., and Zhong Yang, G. (2003) 'The role of bioinformatics in two-dimensional gel electrophoresis', *Proteomics*, 3: 1567–96.

Garfin, D.E. (2003) 'Two-dimensional gel electrophoresis: an overview', *Trends in Analytical Chemistry*, 22: 263–71.

Goar, A., Postel, W., Gunther, S. and Weser, J. (1985) 'Immobilized horizontal two dimensional electrophoresis with hybrid electrophoretic focusing in immobilized pH gradient in the first dimension and laying-on transfer to the second dimension electrophoresis', *Electrophoresis*, 66: 599–604.

Gorg, A., Obermaier, P., Boguth, G., Harder, A., Scheibe, B., Wildgruber, S. and Weiss, W. (2000) 'The current state of two-dimensional electrophoresis with immobilized pH gradients', *Electrophoresis*, 21: 1037–53.

Issaq, H.J. and Veenstra, T.D. (2008) 'Two-dimensional polyacrylamide gel electrophoresis (2D-PAGE): advances and perspectives', *Biotechniques*, 44: 697–700.

Liebler, D.C. (2002) *Introduction to Proteomics: Tools for New Biology*. Totowa, NJ: Humana Press.

Lilley, K.S., Azam, R. and Paul, D. (2002) 'Two-dimensional gel electrophoresis: recent advances in sample preparation, detection and quantification', *Current Opinion in Chemical Biology*, 6: 46–50.

Nugues, P.M. (1993) 'Two-dimensional electrophoresis image interpretation', *IEEE Transactions on Biomedical Engineering*, 40: 760–70.

O'Farrell, P.H. (1975) 'High resolution two-dimensional electrophoresis of proteins', *Journal of Biological Chemistry*, 250: 4007–21.

Pennington, S.R. and Dunn, M.J. (eds) (2001) *Proteomics: From Protein Sequence to Function*. Oxford: BIOS.

Web addresses

http://expasy.org/ch2d/.
http://web.mpüb-berlin.mpg.de/cgi-bin/pdbs/2d-page/extern/references.cgi.
http://www.ncbi.nlm.nih.gov/Class/NAWBIS/Modules/Protein/protein19.html.

12

Mass spectrometry for proteomics

The curiosity remains, though, to grasp more clearly how the same matter, which in physics and in chemistry displays orderly and reproducible and relatively simple properties, arranges itself in the most astounding fashions as soon as it is drawn into the orbit of the living organism.

Max Delbruck, *A Physicist Looks at Biology*, 1949

Abstract: In the genomics and proteomics era, high throughput, sensitive techniques are needed to identify and characterize gene products. Mass spectrometric application for protein analysis was reinvented after the development of soft ionization processes. In this chapter, a brief history of the mass spectrometer is presented with details of basic instrumentation of mass spectrometer. Characterization of proteins using the mass spectrometer is also discussed.

Key words: EI, ESI, FAB, ion trap analyzer, MALDI, mass accuracy, mass range, monoisotopic mass, quadruple analyzer, resolution of mass spectrometer, ToF analyzer.

Key concepts

- The mass spectrometer separates the molecular ions based on their mass to charge (m/z) ratio.
- Mass spectrometers work under vacuum in order to prevent the collision of the ion molecules with air in the instrument.
- Protein sample to be analyzed should be converted into gas phase ions for mass spectrometric analysis.
- Proteins are thermolabile, less volatile and polar and require the soft ionization process.
- The most commonly used soft ionization sources for protein analysis are MALDI and ESI.

12.1 Introduction

The sizes of the molecules vary from small to large biological macromolecules such as proteins, carbohydrates and nucleic acids. Different atoms combine in different ways to produce different molecules. Therefore, each molecule will have a unique molecular mass, for example, the molecular mass of CH_4 is 13 and H_2 is 2. Therefore, the molecular mass of each group helps to identify the type of molecules in the unknown sample. Large samples like proteins or carbohydrates are further broken down and an accurate mass of the constituent fragmented molecules is determined using a mass spectrometer. From the mass spectrum, we can deduce the structure of the molecule. The above concept can be practically done using mass spectrometry.

Mass spectrometry (MS) is an indispensable high throughput proteomics technique. A mass spectrometer is an instrument capable of the ionizing the sample, separating them based on their mass to charge ratio and recording the mass to charge ratio and relative intensities of ions to produce a mass spectrum. It operates under specified vacuum conditions. Initially, MS was used for physicochemical analysis of hard samples. Thanks to soft ionization techniques like **ESI** and **MALDI**, which enable the use of mass spectrometric analysis of biological molecules like proteins, carbohydrates and nucleic acids, in proteomics, mass spectrometry is being used for protein identification, quantification and sequencing, location and identification of post-translational modifications.

Due to the invention of the mass spectrometer, many breakthroughs have been made in physics and chemistry, such as the exact determination of atomic mass, the characterization of new elements, the determination of trace pollutants, etc. Development of biological mass spectrometry has been a boon to proteomics scientists. Mass spectrometric analysis of biomolecules needed two improvements over the traditional MS instruments, namely increased mass range suitable for high molecular weight analysis, and soft ionization techniques to work with thermolabile biomolecules. One of the limitations of mass spectrometric analysis is that it is a destructive method, as a sample cannot be recovered after analysis.

12.2 History of the mass spectrometer

Before the 1960s, mass spectrometric analysis was always associated with physical and chemical analysis of samples but many improvements were made for biomolecule analysis. In the early twentieth century, many scientists in Europe were researching on measuring the mass of the atomic particle, specifically cathode rays. In 1897, J.J. Thomson joined as a Cavendish Fellow at Cavendish Laboratories, Cambridge University, and started working on cathode rays. During the First World War, F. Aston demonstrated the existence of isotopes of non-radioactive elements. He also succeeded in separating the isotopic elements based on their mass difference using the electrostatic magnetic field and described the first mass spectrometer. The first developed mass spectrometer was very simple; a gas discharge tube was used to generate gas phase ions and exposed to an electric and magnetic field. The gas phase ions were deflected onto a photographic plate. He won the Nobel Prize in 1922 for his pioneering work on mass spectrometry. In 1886, E. Goldstein described the existence of positive rays in the discharge tube moving opposite to the already known cathode rays. Joseph John Thomson (1906) proved that the gas molecules can conduct electricity.

Real interest in the mass spectrometer was developed again during the Second World War, as it was discovered that uranium has a fission property and is capable of releasing an enormous amount of energy. Many famous scientists wondered which isotope of uranium was responsible for radioactivity. Mass spectrometric analysis revealed that ^{235}U is responsible for radioactivity, and then nuclear gas was developed from ^{235}U. Until the 1940s, the mass spectrometer was used as only as a quantitative tool for chemists; because of its huge investment costs and limited application, the mass spectrometer did not find a place in academic laboratories.

In the following years, many mass spectrometers were developed with an improved ionization source, a mass analyzer and detectors with high resolving power. In 1946, W. Stephens reported a concept in mass spectrometry with a new type of analyzer called a Time of Flight (**ToF**). In 1968, M. Dole developed a softer ionization technique called the electrospray ionization (ESI) technique. The following scientists were awarded a Nobel Prize for their contributions to improvements in the MS. Wolfgang (1989) for the ion trap analyzer, John Bennet Fenn (2002) for the ESI technique and Kochi and Tanaka (2002) for MALDI. Nowadays, mass spectrometry has become an integral part of any research. According to the IUPAC definition, mass spectrometry is the branch of science dealing with aspects of mass spectroscopes and the results obtained with these instruments.

12.3 Mass spectrometer

12.3.1 Basic MS instrumentation

A mass spectrometer consists of three important parts, namely, the ionization source, the mass analyzer and the ion detector. The mass spectrometer measures the mass to charge ratio of the gas phase molecules. According to IUPAC, m/z is a dimensionless quantity and it is obtained by dividing the mass number (m) by its charge number (z), for example, the m/z ratio of $C_7H_7^{2+}$ is 45.55. When the charged particles are passed through a magnetic field, the particles are deflected, and usually they take a circular path. The degree of deflection depends on the mass of the particle being deflected. Therefore, the basic requirement for mass spectrometric analysis is that the particle should be in its charged state. Uncharged molecules cannot be analyzed using a mass spectrometer as only the charged particles are deflected by the magnetic field.

The first stage of mass spectrometry is the ionization source. The sample is placed inside the ionization chamber and different methods of ionization (discussed under MS instrumentation types) are employed to ionize the sample molecules and convert them to the gas phase. The molecular ions can be created by either the addition or removal of the electron/proton. The gas phase molecular ions are pulled or accelerated towards the second stage of the MS instrument called the analyzer or the mass analyzer. The function of the mass analyzer is to separate the mixture of the molecular gas phase ions received from the ionization source. The separation is based on the mass to charge ratio of the charged molecule. The separated ions are detected using a detector to record the mass to charge ratio and intensity of ions. All four sectors in the MS instrument (see Figure 12.1) are interlinked and are controlled by a computer.

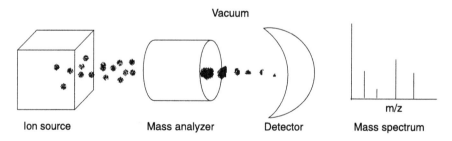

Figure 12.1 A mass spectrometer. It has three compartments: the ion source where the sample is ionized, the mass analyzer separates the gas phase ions from the ionization source according to the mass to charge ratio, the detector records the intensity of ions and the mass spectrum is generated. The entire set-up is maintained under high vacuum conditions.

Published by Woodhead Publishing Limited

The vacuum in a mass spectrometer

A mass spectrometer is maintained under high vacuum (10^{-6}–10^{-9} Torr). The vacuum system maintained in the MS instrument ensures that the charged molecules do not interfere with the air molecules. Ions react fast to form more stable products. Their interaction with other molecules has to be avoided. It increases the chance of molecules reaching the detector without losing any energy.

12.3.2 Ionization sources

This is the first section of the mass spectrometer, maintained at relatively low vacuum (10^{-3} Torr). The sample is introduced into the ionization source either in liquid or gaseous or solid state. For biomolecular characterization, usually a liquid or solid state is used. The most commonly used ionization mode is the positive ion mode. In this method, the sample is introduced into the ionization chamber and a high energy electron beam is passed perpendicular to the sample direction. Interaction between the electron beam and sample molecule results in the removal of the outer electron, thereby a positively charged molecule is produced. Singly charged ions are produced, as it is difficult to remove the second electron from the already ionized molecule. Sometimes if the sample is electronegative in nature, it captures the electron and becomes a negatively charged ion. In some cases, if the sample is inert so that it is not easily ionizable, additional ions can be added to make the sample molecule ionized. This kind of ionization is known as adduct ionization and is based on the nature of the sample being studied. The two main types of ionization sources are MALDI and electrospray ionization.

Ionization sources

1. Chemical impact ionization (CI).
2. Thermal ionization (TI).
3. Field ionization.
4. Electron impact ionization (EI).
5. Electrospray ionization (ESI).
6. FAB.
7. MALDI.

Matrix assisted desorption and ionization (MALDI)

MALDI is a soft ionization technique, specifically developed for biomolecule analysis. A nitrogen laser beam is used to excite matrix chemicals. The matrix chemicals are small organic molecules with chromophores which absorb a laser beam and transfer energy to the co-crystallized biomolecules, thereby preventing the destruction of the biomolecule by the laser energy. As the photons from the small molecules are transferred, it helps the protein to ionize and evaporate into the gas phase. The matrix chemicals are 3, 5 dimethoxy 4-hydroxycinnamic acids (sinapinic acid), 2,5-dihydroxy benzoic acid, etc. The matrix solution is prepared in an acetonitrile and water mixture. The matrix chemicals are acidic in nature and act as a proton source for ionization. The protein to be analyzed is mixed with the matrix solution and spotted on a solid surface and allowed to vaporize. Then the slide containing the crystals is inserted into the ionization source and the laser beam is projected on each spot in a pulsed manner. The MALDI source (Figure 12.2) works well with larger molecular weight biomolecules and is usually used for mass determination. MALDI also tolerates some amount of salt contamination in the sample. For high throughput analysis, the MALDI sample preparation is completely automated.

Electrospray ionization (ESI)

ESI source is another important soft ionization method used for biomolecule analysis (Figure 12.3). In this method, liquid samples are used. Modern MS

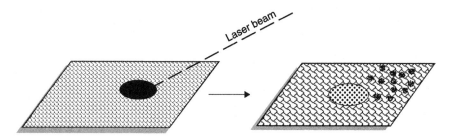

Figure 12.2 MALDI ionization source. Purified peptide is mixed with matrix chemicals and spotted on the solid probe to form co-crystals and then exposed to pulsed laser beam. The matrix chemicals absorb laser energy and transfer the photons to peptides which results in ionization and sublimation of the peptides.

Published by Woodhead Publishing Limited

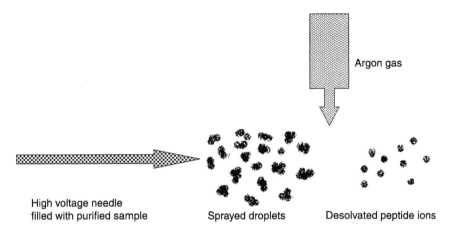

Argon gas

High voltage needle
filled with purified sample   Sprayed droplets   Desolvated peptide ions

Figure 12.3 Electrospray ionization source. The purified protein sample is passed through a fine tip needle which is maintained at a high voltage and the sample is sprayed as droplets with solvents. Argon gas is sprayed on these droplets to form desolvated gas phase peptide ions.

instruments are integrated with liquid chromatography. Peptide fragments after trypsin digestion and are purified using liquid chromatography. The purified peptides are sent to the ESI source of the mass spectrometer. The sample is sprayed through a narrow capillary needle which is maintained at a high voltage. The sample is sprayed as fine droplets. When a dry inert air like argon is passed over the droplets, the mobile phase evaporates, rendering formation of the ions. Usually the ESI source produces multiply charged ions.

12.3.3 Mass analyzers

This is the second part of the mass spectrometer. It separates the ions based on their mass to charge ratio. Resolution of the mass spectrometer depends on the mass analyzer. There are many types of mass analyzers, namely time of flight, *Quadrupole*, ion trap, magnetic sector, etc. Depending on the number of mass analyzers connected, a particular mass spectrometer can be called an MS or a tandem MS or MS/MS. The ToF analyzer is a cheap and robust mass analyzer and it is used for molecular weight determination. The ion trap is a costly mass analyzer, but a single ion trap analyzer will perform both MS and MS/MS. For structure elucidation, a particular peptide has to be further fragmented.

Types of mass analyzers

1. Time of flight (ToF).
2. Quadrupole.
3. Ion trap.
4. Magnetic sector.

12.3.4 Output of the mass spectrometer

The output of a mass spectrometer is a graph representing bell-shaped peaks. The sharp rise of the peak is due to the increase in the number of ions hitting which causes the corresponding increase in the current in the detector. The strongest ion peak is denoted as the base peak (Figure 12.4).

The ultimate aim of mass spectrometric analysis of a protein or peptide is the accurate determination of mass based on the mass to charge ratio. Therefore, m/z detected from the MS spectrum has to be converted to the mass of the protein or peptide and this is done by dividing the mass of the

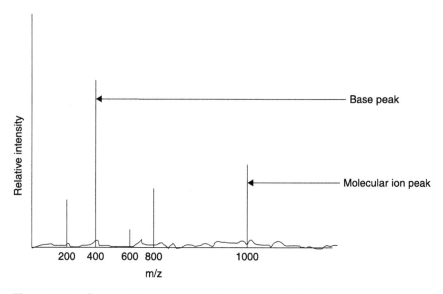

Figure 12.4 Output of the mass spectrometric analysis. The mass spectrum is represented in the form of a graph in which m/z is the x-axis and the relative intensity of peak is on the y-axis.

molecular ion by the charge it carries. Most of the time 'z' will be 1 as mass spectrometric analysis is carried out with singly charged ions and the m/z value directly yields the molecular mass or atomic mass. Electrospray ionization of the proteins or peptides commonly yields multiply charged ions. As the charge increases, the mass to charge ratio of the ion is reduced, or if a particular ion carries twice the charge, its mass will be reduced by half.

12.3.5 Resolution of the mass spectrometer

This denotes the capacity of the mass spectrometer to give an accurate mass measurement of a given molecule. The resolution is calculated by dividing the mass by the peak width. Resolution of the mass spectrometer is defined as the ability of the mass spectrometer to separate ions with close mass by charge ratio. If the resolving power is good, two ions having a closely related mass to charge ratio will be seen as two separate peaks, otherwise two peaks will be overlapping and this limits the accurate mass determination. Resolving power of mass spectrometer is expressed as ppm (part per million).

12.4 Protein sample preparation for MS analysis

Mass spectrometry requires an extremely small quantity of highly purified sample. The quantity of the sample introduced inside the mass spectrometer is very small. It requires a femtomole quantity. If the band is visible with the CBB in SDS-PAGE gel, there is enough for protein identification using MS analysis. It is always good to use spectroscopic grade solvents for sample solubilization. For sample preparation, use of large polar solvents and DMSO should be avoided. It is better to prepare a sample with methanol or acetonitrile. Impurities should be avoided as they compete with analytes for ionization and also show additional peaks in the mass spectrum. Salts are detrimental to MS analysis. Usually they form adduct peaks in the graph and compete with the molecular ion peak. The tolerable salt concentration in the mass spectrometric sample is less than 1 mM. The presence of sodium and potassium ions in the sample causes real problems in the MS analysis. The presence of ammonium acetate ions up to 20 mM is tolerable. The personnel involved in MS studies should use gloves to avoid keratin contamination. The sample preparation should be carried out in a dust-free room. It is recommended that high quality reagents and milliQ water be used to ensure there is no salt or unwanted protein contamination. It is always recommended to use siliconized polypropylene tubes and non-retention tips to avoid protein loss. The sample can be introduced into the ionization source of the MS instrument in any state such as solid, liquid or gas, but it should be converted to the gas phase in the ionization chamber and sent to the mass analyzer. Proteins are large molecules and they cannot be analyzed in a mass spectrometer directly. They must be cleaved into small peptides using sequencing grade endoproteases. Some of the commonly used sequencing grade endopeptideases are trypsin, Lys-C, Asp-N and Arg-C.

12.5 Applications of MS proteomics

The first developed mass spectrometers had low *mass range* and applied harsh conditions for ionization and vaporization. These conditions are not suitable for biomolecular applications. Many modifications were made to ionization mechanisms and vaporization that have made MS suitable for biomolecular analysis. Mass spectrometry has become a versatile high throughput technique for proteomics. It is used for protein identification, protein quantification, protein sequencing, and post-translational modification determination.

Applications of mass spectrometry

1. *Molecular weight determination.* It is possible to determine accurate molecular weight of the peptides using MALDI–ToF MS.
2. *Protein sequencing.* Proteins are fragmented to peptides by the endoprotease treatment and the peptides are further fragmented in tandem MS. From the mass spectrum, amino acid sequence can be determined.
3. *Protein identification.* Purified protein is fragmented into small peptides by treating them with trypsin. The peptides generated form the fingerprint of that protein. Using MALDI-ToF, the accurate mass of these peptides is determined. Comparing the partial mass list with the database, the unknown protein is identified. This is known as peptide mass fingerprinting.
4. *Glycoprotein analysis.* Tandem mass spectrometry is used for glycoprotein analysis. The glycosylated proteins are enriched and

---

subjected to fragmentation. Using mass spectrum data, it is possible to identify the type of glycan and site of glycosylation.

5. *Phosphoprotein analysis*. Tandem mass spectrometry is used for phosphoprotein analysis. The phosphorylated proteins are enriched and subjected to fragmentation. Using mass spectrum data, it is possible to identify the site of phosphorylation.

6. *Protein quantification*. Expression profiling can be performed using mass spectrometric data. Proteins are labelled differentially either prior to isolation (SILAC) or after isolation (ICAT). The protein samples are labelled with a light isotope and a heavy isotope, mixed and separated in 2D-PAGE. The selected protein spots are excised from 2D-PAGE and subjected to mass spectrometric analysis. Based on the intensity of the peaks, which differ by one mass unit, the fold expression of those proteins can be calculated.

Review questions and answers

1. *What is the effect of a number of charges on a molecular ion separation in mass spectrometric separation?*

As the charge increases, the mass to charge ratio of the ion will be reduced, otherwise if a particular ion carries twice the charge, its mass/charge ratio will be reduced by half.

2. *Why is it necessary to have a free path or vacuum in mass spectrometers?*

It is important to maintain a high vacuum in the mass spectrometer. This will avoid collision of the ion with air molecules and prevent the molecular ions proceeding towards the path of the mass spectrometer.

3. *What is an adduct ion? When would you use an adduct for the ionization process?*

According to the IUPAC definition, adduct ions are ions containing all the constituent atoms of one species as well as an additional atom or atoms. Usually, MS analysis of carbohydrate sodium ions are used as an adduct ion to provide a charge to the carbohydrates.

Published by Woodhead Publishing Limited

4. *Conventional mass spectrometers are not suitable for biomolecular analysis. Why?*

Conventional mass spectrometers were developed for physical and chemical characterization of hard elements involving harsh ionization procedures like thermal ionization or electron impact ionization which are not suitable for soft biomolecular analysis.

5. *What are molecular ions and fragment ions?*

During the ionization process of the mass spectrometric analysis of biomolecules, an electron is removed or a proton is added to the organic molecule to form an ion, this is known as a molecular ion. This molecular ion is unstable and further fragmented to produce smaller ions, and these ions are called fragment ions. In the mass spectrum, the molecular ion will be the heaviest and the fragment ions will be lighter.

6. *What do we mean by the mass range of MS instrument?*

This is the maximum mass to charge ratio of ions that can be determined using a mass spectrometer.

7. *Accuracy of mass determination using MS data depends on the resolving power of the MS instrument. Justify this statement.*

The resolving power of the mass spectrometer is defined as the ability of the mass spectrometer to differentiate ions with close mass to charge ratios. If the resolving power of the mass spectrometer is high, the mass of the ion can be the monoisotopic mass of the ion, which is the accurate mass of the ion.

8. *List the applications of mass spectrometric analysis in proteomics field.*

- Accurate determination of the molecular proteins.
- Protein identification using peptide mass fingerprinting.
- *De novo* sequencing of peptide and protein identification.

Recommended reading

Burlingam, A.I. (ed.) (2005) *Biological Mass Spectrometry in Methods in Enzymology*, Vol. 402. Palo Alto, CA: Elsevier Academic Press.

Finehout, E.J. and Lee, K.H. (2004) 'An introduction to mass spectrometric application in biological research', *Biochemistry and Molecular Biology Education*, 32: 93–100.

Griffith, J. (2008) 'A brief history of mass spectrometry', *Analytical Chemistry*, 80: 5678–683.

Jonsson, A.P. (2001) 'Mass spectrometry for protein and peptide characterization', *Cellular and Molecular Life Sciences*, 58: 868–84.

Published by Woodhead Publishing Limited

13

Protein Identification by Peptide Mass Fingerprinting (PMF)

Abstract: Peptide mass fingerprinting is a high throughput protein identification technique in which an unknown protein is digested with endoprotease to yield the constituent small peptides. The accurate mass of these peptides is determined by MS analysis. This gives the peak list of peptides of the unknown protein. This peak list is compared with the theoretical peptide peak list obtained from the *in silico* digestion of the database proteins and the best match is identified by computer software. The main advantage of this method is that it does not depend on protein sequencing for protein identification. The limitation of this method is that it requires the database to have the protein which is already characterized on another organism.

Key words: endoproteases, *in silico* digestion, monoisotopic mass, peak list, PMF, protein database.

Key concepts

- Proteins are ultimately known by their sequence.
- Protein identification using sequencing is time-consuming.
- Proteins can be identified based on the peptide mass fingerprint which is known as peptide mass fingerprinting (PMF).
- In PMF, protein identification is performed by comparing the mass spectrometric peak list with that of the theoretical peak list generated from databases.

13.1 Introduction

Protein identification is an important step in the characterization of a protein for further studies. Although proteins are separated by 2D-PAGE which reveals some parameters about the proteins such as pI and molecular weight, theses two criteria are not sufficient to describe a protein completely. The ultimate way of describing any protein is by its sequence as that is unique to a particular protein. Protein identification is done after sequencing by **Edman degradation** but this is a time-consuming method and also limited by the size of the protein sequence that can be determined. A faster method of protein identification was needed to cope with genome sequencing projects. Application of mass spectrometric data for protein identification was tested using the **fast atom bombardment (FAB)** mass spectrum of trypsin digested lysozyme. The lysozyme was successfully identified by correlating the measured peptide mass with the already existing peptide mass of lysozyme by using a computer program called FRAGFIT. Due to the lack of sensitivity of FAB, mass spectrometric data was not utilized for other studies. MALDI and ESI mass spectrometric techniques were introduced in 1992. These techniques are sensitive, work with the femtomole quantity of the sample and the mass range is up to 200 kDa. They became a handy tool for peptide mass determination. Mass spectrometry replaced the cumbersome protein chemistry-based protein identification. MALDI-ToF is commonly used mass spectrometry for protein identification by PMF, a rapid and sensitive technique for protein molecular mass determination. The success of identifying a correct protein depends on the accuracy of the mass of the peptides that is used for comparison. It is shown that mass accuracy of 50 ppm gives a good result for protein identification by PMF. Another important point for successful protein identification by PMF is correct enzyme specificity.

13.2 Principles of peptide mass fingerprinting

The total number of protein entries in Uniprot was more than 523,151 up to the end of November 2010. Digestion of a protein with a specific **endoprotease** yields a defined number of peptides of specific length and mass. These peptides with specific masses are unique to a protein, therefore, this is called the peptide mass fingerprint of that protein. In **peptide mass fingerprinting,** all the database protein sequences are downloaded and subjected to virtual digestion with a selected endoprotease. This process is known as *in silico* **digestion,** and using this process a theoretical **peak list** is prepared in which the virtual peptides are in descending order of their mass. This serves as the reference peak list for peptide mass fingerprinting. Therefore, a protein in the database is digested into small peptides and represented in the peak list. The original protein can be identified from the

Published by Woodhead Publishing Limited

peptide masses. Similarly, an unknown protein can be identified if the peptide mass list of this protein matches the peptide mass list of a database protein. The success of this method depends on mass accuracy and the number of peptide masses compared for each protein.

Steps in peptide mass fingerprinting

1. Isolation of intact protein by 1D- or 2D-PAGE.
2. Digestion of protein into small peptides by proteolytic treatment with trypsin.
3. Analysis of peptides using mass spectrometry.
4. Preparation of peak list from mass spectrum.
5. *In silico* digestion of database proteins and generation of theoretical peak list.
6. Comparison of peak list and theoretical peak list to get best match.

13.3 Protein preparation for PMF

Biological sources have thousands of proteins. In order to isolate a single protein from the total proteins in the source material, the total protein is subjected to gel electrophoresis. The commonly used proteomics method is 2D-PAGE as it can separate thousands of protein in a single gel based on their pI and molecular weight. The proteins separated on 2D-PAGE gel are excised from the gel and used for proteolytic digestion. The 2D-PAGE gel has to be washed thoroughly before excising the spot. A sterile unused scalpel with a fine tip is used to remove the protein spot. The protein present inside the SDS-PAGE gel is subjected to proteolytic digestion. Before proceeding to digestion, the gel has to be washed thoroughly to remove water and buffer. Commonly used proteases for peptide mass fingerprinting are trypsin chymotrypsin, LysC, AspN, ArgC, etc. These proteins are specific endoproteases and their cleavage specificities are given in Table 13.1. Since these proteases cleave proteins at particular points after complete digestion, each protein will yield a defined number of short peptides with characteristic mass decided by the amino acid and constituent post-translational modifications, if any. The gel piece is dried by mixing with ammonium bicarbonate and an acetonitrile solution and evaporating it in the speedvac. The dry gel pieces thus obtained are re-suspended in a trypsin solution and

Table 13.1 Specific proteases for protein digestion and their properties

Sl. No.	Name of the enzyme	Cleavage site	Source	Cutting site
1	Trypsin	Arginine (R) or Lysine (L) But not followed by Proline (P)	Bovine pancreas	C-terminal
2	Chymotrypsin	Phenylalanine, tyrosine, Trypsine	Bovine pancreas	C-terminal
3	Lys-C	lysine	*Lysobacter enzymogenes*	C-terminal
4	Glu-C	After Glutamate	*S. aureus*	C-terminal
5	Asp-N	Before Aspartate	*Pseudomonas fragi*	N-terminal
6	Pepsin	Phenylalanine, tyrosine, Trypsine	Porcine stomach	N-terminal
	S. aureus V8 protease	Aspartic acid, Glutamic acid	*S. aureus*	C-terminal

incubated for 16 hours at 37°C. Care should be taken not to include excess trypsin for protein digestion as it undergoes autolysis and produces additional peaks. The accepted ratio of sample to trypsin is 50:1. After complete digestion, the protein is centrifuged and re-suspended in 1 per cent trifluroacetic acid. Incubation time should also be optimized to obtain the correct result.

13.4 Mass spectrometric analysis of peptide fragments

High quality chemicals are required for mass spectrometric work. For peptide mass fingerprinting, MALDI-ToF works better than ESI-MS. A small quantity (a picolitre) of the purified peptide sample is mixed with matrix chemicals to form crystals and deposited on probe surface. Upon drying, the peptides co-crystallize with the matrix chemicals and the target plate is inserted into the ionization sources of the MALDI-MS instrument. The output of the mass spectrometer provides a list of peaks which corresponds to the individual peptides derived from digestion of a protein.

13.5 Data analysis and identification of protein

The peak list is compared with a peak list generated from the database proteins. The commonly used computer search engines are MS-Fit, Mascot,

Peptident, Profound, etc. The monoisotopic mass of the each peak, the protease used, the number of missed cleavages in order to account for the possible incomplete digestion are given as input. It is better to limit the comparison to be made by selecting related organisms and a selected database to get a quick result. Information such as pI and Mr of the protein has to be provided. Additional information required is the modifications of the amino acid side chains such as cysteine oxidation, tyrosine phosphorylation to get a correct estimate of the peptide mass. It is also important to provide the number of peaks to be compared with the *in silico* peak list. The default setting is four peaks, and then an algorithm matches the four peak masses with the *in silico* peak masses. A protein hit showing three peaks matched will be discarded. Mass tolerance can be set to find the best match (Figure 13.1).

Protein identification by mass spectrometry replaced the laborious protein sequencing and identification. This technique increased the speed and sensitivity of peptide sequence many fold. Protein identification depends on the accuracy of mass to charge ratio measured. The number and size of the peptide fragments depend on proteolysis. The most commonly used technique for peptide mass fingerprinting and protein identification is MALDI mass spectrometer. Figure 13.2 shows the flow chart in peptide mass fingerprinting.

13.5.1 ALDENTE peptide mass fingerprinting tool

ALDENTE is a powerful PMF tool maintained by the ExPASy proteomics server of the SWISS Institute of Bioinformatics. The pI (Min. 0 to 14) and MW of the proteins have to be supplied along with other information such as type of protease. Commonly used proteases are trypsin, chymotripsin, ArgC, ASP N, Proteinase K, etc. Probable missed cleavages also are selected to account for the incomplete digestion. Possible post-translational modifications and experimental modifications are also submitted to get the best match from the database.

Review question and answer

1. *Protein identification by 2D-PAGE is not sufficient. Why?*

2D-PAGE separates proteins based on pI and molecular weight of proteins. Because more than one protein will have the same pI and molecular weight, the 2D-PAGE result is not sufficient. The ultimate characterization of proteins is done by its amino acid sequence.

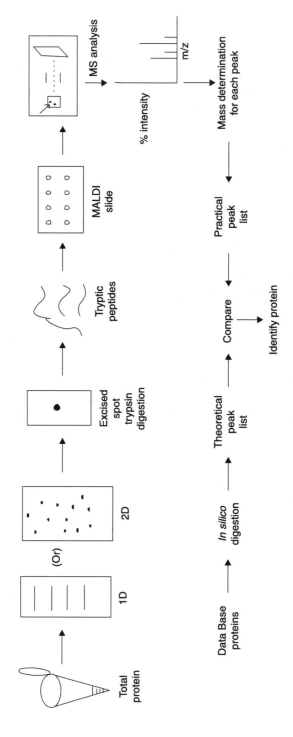

Figure 13.1 Steps involved in PMF. Total protein isolated from the sample is separated on polyacrylamide gel electrophoresis. A protein spot is selected and excised from the gel and subjected to endoprotease digestion to get peptides and sent for mass spectrometric analysis. From the mass spectrum, an experimental peak list is prepared. From the database, protein sequences are downloaded and *in silico* digestion is performed and theoretical peak list is prepared. Using computer software, a comparison is made between the experimental and the theoretical peak list. The output will give the matching hits which are arranged according to their matching score.

Published by Woodhead Publishing Limited

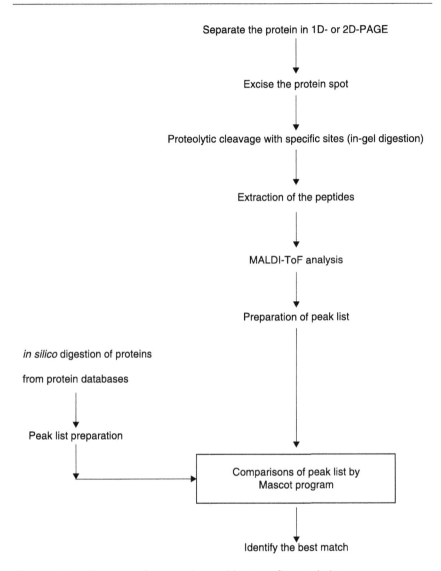

Figure 13.2 Flow chart for steps in peptide mass fingerprinting.

Recommended reading

Gevaert, K. and Vandekerckhove, J. (2000) 'Protein identification methods in proteomics', *Electrophoresis*, 21: 1145–54.

Henzel, W.J., Billeci, T.M., Stults, J.T., Wong, S.C., Grimley, C. and Watanabe, C. (1993) 'Identifying proteins from two-dimensional gels by molecular mass searching of peptide fragments in protein sequence database', *Proceedings of National Academy of Sciences, USA*, 90(11): 5011–15.

James, P., Quadroni, M., Carafoli, E. and Gonnet, G. (1993) 'Protein identification by mass profile fingerprinting', *Biochem. Biophys. Res. Commun,* 195(1): 58–64.

Madin, K., Hermann, S., Kirr, L., Schmitt, D. and Koch, T. (2007) 'Enzymatic: protein digest for mass spectrometric analysis', *Biochemica,* 3: 23–25.

Mann, M., Højrup, P. and Roepstorff, P. (1993) 'Use of mass spectrometric molecular weight information to identify proteins in sequence databases', *Biol. Mass Spectrom,* 22(6): 338–45.

Pappin, D.J., Højrup, P. and Bleasby, A.J. (2006) 'Rapid identification of proteins by peptide mass fingerprinting', *Current Biology,* 3: 327–32.

Webster, J. and Oxler, D. (2005) 'Peptide mass fingerprinting; protein identification using MALDI-ToF mass spectrometry', *Methods in Molecular Biology,* 310: 227–40.

Web address

http://expasy.org/tools/aldente/.

14

Protein sequencing techniques

It is certain that proteins are extremely complex molecules but they are no longer beyond the reach of the chemist.

— F. Sanger

Abstract: The ultimate structure and function relationship of any protein lies in its amino acid sequence. The order of the amino acid sequence of a protein is called the primary structure of the protein. Protein sequencing can be done by three methods: Edman degradation, tandem mass spectrometry, and the bioinformatics method in which the protein sequence is derived from cDNA or mRNA or a DNA sequence available in databases. In this chapter, protein sequencing methods such as Edman degradation and tandem mass spectrometry are discussed in detail.

Key words: *de novo* sequencing, Edman degradation, tandem mass spectrometry.

Key concepts

- Proteins are made up of twenty different amino acids which are arranged in a unique order in each protein. This is known as the primary structure of a protein.
- Modern protein sequencing is done by using automated protein sequencers which are based on the principle of Edman degradation.
- Tandem mass spectrometry is also used to sequence the proteins and has many advantages over Edman sequencing.
- For protein sequencing, proteins have to be purified to homogeneity using liquid chromatography.
- Intact proteins cannot be sequenced using a mass spectrometer. Purified protein is treated with trypsin to convert it into peptides of a small size suitable for mass spectrometric analysis.

Published by Woodhead Publishing Limited

- The Edman degradation method has some limitations and is one of the important methods for protein sequencing. Mass spectrometry has become an alternative to the Edman degradation method where N-terminally modified amino acid cannot be modified.

14.1 Introduction

Proteins are complex biological macromolecules and they are synthesized in the cells by the process of translation. The sequence of the amino acids in a protein is determined by the sequence of bases in the mRNA. The sequence of the mRNA in turn is determined by the gene sequence present in the genome. After synthesis, the protein undergoes folding and other post-translational modifications. The native protein has secondary structures and tertiary structures which are required for its biological activity. Many biologically active proteins are made up of more than one polypeptide chain. The chemical structure of proteins was unknown for many years, till Sanger successfully determined the sequence of two polypeptides of bovine insulin in 1953. He used approximately 100 mg of protein and he worked for about 10 years on it. It was known that proteins have different amino acid composition, but only after Sanger's sequencing of insulin it was established that each protein has a unique sequence and that determines its structure and its functions. Protein sequencing reveals the order of amino acids in a polypeptide chain.

There are three major techniques for protein sequence determination. They are **Edman degradation, mass spectrometry** and amino acid sequence prediction from the cDNA or DNA sequence from the databases.Edman degradation is based on N-terminal modification followed by cleavage of one amino acid at a time from the rest of polypeptide chain and then identifying the modified amino acid that is released by chromatography. It is currently followed to determine the sequence of a large number of unknown proteins. Thanks to the new sensitive mass spectrometric techniques like ESI-MS in tandem mode to achieve the fragmentation, the amino acid sequence can be identified from the mass spectrum using computer algorithms. This technique has become popular for protein sequence determination followed by protein identification from database.

Three methods to determine the protein sequence

1. Edman degradation.
2. *De novo* sequencing by mass spectrometry.

Published by Woodhead Publishing Limited

3. Reverse translation of gene sequence to protein sequence from database.

14.2 Preparation of protein sample for sequencing

The protein sample to be sequenced may be in a liquid state or a solid state. Proteins that are **electro-blotted** on PVDF membranes are commonly used. The amount of sample required for sequencing is usually small (less than 1ng) in the range of 10 to 100 picomoles. The liquid sample should be solubilized in a volatile solvent such as acetonitrile or propanol. The samples can be supplied in lyophilized form, but it is preferable for it to be on the PVDF membrane. The liquid sample may contain salts, SDS, buffers, or free amino acids which will contaminate the sequencing machine. In order to identify cysteine, it should be alkylated. The sample should not be blotted on a nitrocellulose membrane.

The sample should be free from any primary amine such as tris or glycine. The salt concentration should be less than 2 mM. The sample should be desalted by dialysis, or gel filtration chromatography. To remove the glycine used during SDS-PAGE, the proteins must be transferred to the PVDF membrane by electro-blotting. Since the PVDF membrane is resistant to **Edman reagent**, it can be used in a protein sequencer, but not a nitrocellulose membrane. After electro-blotting, the membrane should be washed with water 4 or 5 times for the complete removal of the tris or glycine from the blot. These primary amines will create many baseline peaks which will interfere with amino acid identification. It is important that the sample is blotted on the membrane and the electro-blot can be stained with CBB. Approximately 40 mm² PVDF membrane with visible staining is sufficient to produce the protein sequence. In general, a quantity of 5 pmol of the sample is sufficient to give a reliable result.

14.3 Steps in protein sequencing

14.3.1 Determination of amino acid composition

This is done prior to sequencing. A known quantity of protein is completely hydrolyzed into its constituent amino acids, and these are quantified to find the relative proportion of each amino acid in the protein. This will help us to discover the ambiguities that may arise during sequencing. This is done by hydrolyzing the protein in the presence of 6 M hydrochloric acid for 24 hours at 100°C. Certain amino acids are degraded during this acid hydrolysis process. In order to make an account of these amino acids, samples are removed and analyzed at different time intervals and the result is extrapolated for complete hydrolysis. The hydrolyzed amino acids are separated using chromatographic techniques and the amount of each amino acid is estimated using the ninhydrin method and absorbance is measured to correlate with the quantity of amino acids.

14.3.2 N-terminal amino acid determination

It is necessary to identify the N-terminal amino acid before starting to analyze the protein sequence. Information about the number of polypeptides in the mixture can be gathered based on the number of different amino acids in that polypeptide. N-terminal sequencing can be done in three ways: (1) Sanger's method; (2) fluoro 2,4-dinitrobenzene (FDBN); and (3) the dansyl chloride method.

14.3.3 C-terminal sequencing

This is done with an enzyme known as carboxypeptidase. It specifically cleaves the amino acids from the C-terminal end of the polypeptide.

14.3.4 Disulfide bond breakage

Since the disulfide bonds between two cysteine amino acids interfere with protein sequencing, they must be reduced. Usually this is done by the reducing agent beta mercaptoethanol followed by iodoacetic acid to convert the reduced SH group of the disulfide bonds. It can also be oxidized to cysteic acid using performic acid.

14.3.5 Proteolytic cleavage of the proteins

Sequencing cannot be done with large polypeptides. They must be cleaved with some proteases to obtain the smaller peptides (15–25 amino acids) that are suitable for sequencing. Usually it is done specifically using specific proteases.

Published by Woodhead Publishing Limited

14.4 Protein sequencing by Edman degradation

This method was developed by Pher Edman in 1960 as an alternative to Sanger's method of protein sequencing. To sequence the proteins used in this method, the proteins are first separated on SDS-PAGE gel and blotted on to a solid support like PVDF membrane. The proteins attached to the PVDF membrane are introduced into the automatic protein sequencer. When the proteins on the solid support are exposed to the Edman reagent, N-terminal amino acid is modified. The modified N-terminal amino acid is hydrolyzed under an acidic condition and identified using chromatography. The major breakthrough in protein sequencing came after this because it does not require a new slot of protein for sequencing, the same polypeptide chain which is reduced by one amino acid from the N-terminal end is used again. Automated amino acid sequencers are fabricated based on this method. The biggest disadvantage of this method is that it cannot be used for polypeptide sequencing whose N-terminal amino acid is modified. Figure 14.1 shows a flow chart for Edman degradation.

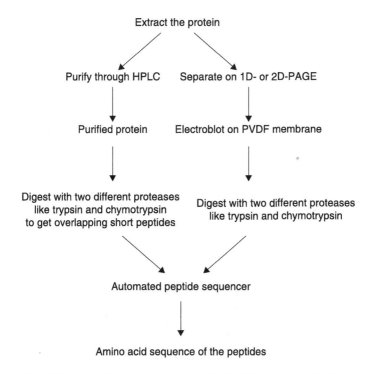

Figure 14.1 Flow chart for protein sequencing by Edman degradation.

Published by Woodhead Publishing Limited

Advantages of Edman degradation

1. Sensitive method, i.e. a picogram quantity of the sample is enough.
2. Intact proteins can be directly analyzed, no need for protein digestion, a small quantity of the protein only is needed.
3. More reliable sequencing technique.

Disadvantages of Edman degradation

1. N-terminal modified proteins cannot be sequenced.
2. Proteins stained with silver nitrate cannot be sequenced.
3. Slow processing; only 10 amino acids are detected in 24 hours.
4. Only up to 50 amino acids can be sequenced in a single reaction.

14.5 *De novo* protein sequencing by mass spectrometry

Protein sequencing using a mass spectrometer has become an important high throughput proteomic technique. It is a *de novo* sequencing method involving determination of the amino acid sequence from the mass spectrum. The sample is purified and subjected to proteolytic digestion. The digested peptides are subjected to both MALDI-MS and tandem MS analysis. From the MALDI analysis, the mass of the peptides will be obtained. In the case of tandem mass spectrometric analysis, the peptides are subjected to further fragmentation. The fragmentation patterns of the polypeptide follow characteristic pattern. Most frequent fragmentation occurs in the peptide bond backbone. One parent ion is selected from the first mass analyzer and is sent to the second analyzer, otherwise known as the CID cell, where the selected peptide is further fragmented to produce daughter ions of smaller size. The fragmentation is random, it forms the 'y' ion and 'b' ion series. The daughter ions are separated and the mass spectrum is obtained. From the mass spectrum it is possible to identify the ascending and descending ion series to find out the amino acid sequence. Initially the tandem MS spectrum was complex and it was difficult to interpret manually. Now there are many algorithms that have been developed to automatically interpret the data (Figure 14.2).

Published by Woodhead Publishing Limited

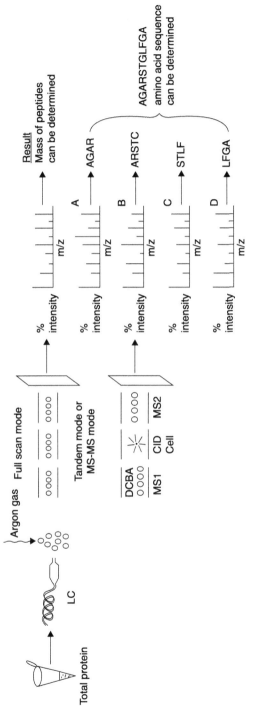

Figure 14.2 Protein sequencing using tandem MS.

> ### Applications of protein sequencing
>
> 1. Structure and function analysis.
> 2. New protein biomarkers can be identified.
> 3. New drug targets can be identified.
> 4. A phylogenetic tree can be constructed to find the evolutionary relationship.
> 5. **Orthologous** and **paralogous** proteins can be identified so that the functions of the genes or proteins can be extrapolated to unknown orthologous proteins.
> 6. Probe sequence for cDNA library screening.
> 7. Protein engineering studies.

Review questions and answers

1. *What are the limitations of the Edman degradation method of protein sequencing?*

The biggest disadvantage of this method is that it cannot be used for sequencing a polypeptide whose N-terminal amino acid is modified.

2. *Why is it preferable to use PVDF electro-blotted proteins for sequencing?*

The samples can be supplied in lyophilized form but it is preferable for them to be on a PVDF membrane. The liquid sample may contain salts, SDS, buffers, free amino acids which will contaminate the sequencing machine.

3. *List the applications of protein sequencing.*

- New protein biomarkers can be identified.
- New drug targets can be identified.
- Phylogenetic trees can be constructed to find the evolutionary relationship.
- Orthologous and paralogous proteins can be identified so that the functions of the genes or proteins can be extrapolated to unknown orthologous proteins.
- Probe sequence for cDNA library screening.
- Protein engineering studies.

4. *Why do we need protedytically cleaved proteins?*

Sequencing cannot be done with large proteins. They must be cleaved with some proteases to get the smaller peptides that are suitable for sequencing.

Recommended reading

Banderia, N., Clauser, K.R. and Pevzner, P.A. (2007) 'Shotgun protein sequencing assembly of peptide from tandem mass spectra from mixture of modified protein', *Molecular and Cellular Proteomics*, 6: 1123–34.

Roepstorff, P. and Fohlman, J. (1984) 'Proposal for a common nomenclature for sequence ions in mass spectra of peptides', *Biomedical Mass Spectrometry*, 11: 601.

Smith, B.J. (2007) *Protein Sequencing Protocols in Methods in Molecular Biology*, Vol. 211, 2nd edn. Totowa, NJ: Humana Press.

Standing, K.G. (2003) 'Peptide and protein *de novo* sequencing by mass spectrometry', *Current Opinion in Structural Biology*, 13: 595–601.

Stretton, A.W. (2002) 'The first sequence: Fred Sanger and insulin', *Genetics*, 162: 527–32.

Zhong, H., Zhang, Y., Wen, Z. and Li, L. (2004) 'Protein sequencing by mass analysis of polypeptide ladders after controlled protein hydrolysis', *Nature Biotechnology*, 22: 1291–6.

Published by Woodhead Publishing Limited

15

Phosphoproteomics

Abstract: Proteins are the most dynamic macromolecules in cells and their structure and function are controlled by post-translational modifications. One of the important post-translational modifications is phosphoproteomics. There are two types of phosphorylation of proteins, such as O-linked phosphorylation and N-linked phosphorylation. Phosphoperoteomics is discussed with respect to their enrichment techniques and mass spectrometric analysis.

Key words: glycoproteins, IMAC, kinases, kinome, phosphatases, phosphoproteins, post-translational modification.

Key concepts

- Most of the proteins are biologically active and perform their function only after post-translational modification.
- Post-translational modifications are either reversible or irreversible.
- Post-translational modifications affect interactions, stability, conformation, localization and functions of proteins.
- Phosphoproteins are an important group of proteins involved in the signal transduction process.
- Phosphorylation and dephosphorylation of proteins are carried out by kinases and phosphatases, respectively.
- The fraction of phosphorylated proteins present in the cell is usually small and cannot be used for analysis directly. Therefore, purification is required, and the most common method is immobilized metal affinity chromatography (IMAC).
- Tandem mass spectrometer is used to find the phosphorylated site in phosphoproteins.

Published by Woodhead Publishing Limited

15.1 Post-translational modifications of proteins

Post-translational modifications (PTM) of proteins play an important role in the cellular functions. PTM is the covalent addition of certain functional groups to the proteins. More than 40 PTMs have been identified and related to diseases such as cancer and neurological disorders. These PTMs play a major role in protein folding, stability, conformation, etc. Of all the post-translational modifications, **phosphorylation** and **glycosylation** are the major players in many of the protein functions. Most proteins undergo some modification before undertaking any function assigned to them. A post-translational modification can be a reversible or an irreversible activity. **Proteolytic cleavage** is one of the common modifications where proteins are cleaved to remove some additional amino acid(s) or portion of protein. Examples are zymogens, which are inactive forms of enzymes and are activated by the removal of some portion of the protein. Methylation is the addition of the methyl group to the lysine side chain responsible for chromatin transcription activity state. Sulfation is a permanent post-translational modification needed for the functioning of the proteins. **Ubiquitination** is another major post-translational modification that has a major role in protein degradation. Small ubiquitin-related modifiers are responsible for SUMOylation. Since PTM occurs at low levels, it is difficult to characterize them using other established proteomics methods. Purification of the post-translational modified proteins is needed. Special detection and purification methods are also needed to study these proteins.

15.2 Phosphoproteomics

Phosphoproteomics is a branch of proteomics in which protein phosphorylation and its associated functions of proteins in a particular cell or tissue or organ are studied. Phosphorylation is the addition of the phosphate group to the side chain of certain amino acid residues in the proteins. Phosphorylation is a sensitive and quick method of altering the cellular functions by altering the activity state of the enzymes either by phosphorylation or dephosphorylation. These are reversible post-translation modifications carried out by the opposite action of two enzymes such as **kinase** and **phosphatase**. More than 10,000 phosphorylation sites have been mapped in the human proteome using cell lines. Phosphorylation is initiated by various external stimuli such as cold, chemicals, heat, etc. and occurs very quickly. A single system will have more than 1,000 kinases, targeting various proteins at various sites. Approximately 30 per cent of the cellular

Published by Woodhead Publishing Limited

proteins are phosphorylated in human proteome. Addition of a phosphate group to the protein increases the polarity of the phosphoprotein making it difficult to ionize during mass spectrometric analysis. Phosphoproteomics studies can either be a global approach involving the study all the phosphorylated amino acids like phosphotyrosine, phosphoserine, phosphothreonine, etc., or targeted approaches involving the study of only one kind of phosphorylation site at a time. Phosphoproteomics study has two objectives: (1) cataloguing of the phosphorylated protein in a particular cell; and (2) assigning functions to the phosphorylated proteins. Protein kinases are the largest gene families identified in humans. They play a vital role in a wide variety of physiological processes and signalling mechanisms.

15.2.1 Effect of phosphorylation on proteins

Phosphorylation of protein at a particular site or more than one site effects many changes in a protein. They are conformational changes that affect the protein's susceptibility to the surrounding environment, localization, protein activity, modification of charge of the protein and protein–protein interaction, and signal transduction. Phosphorylation plays a major role in cell signalling, the regulatory activities of cell division, apoptosis, etc. During phosphorylation, a phosphate group is attached to the groups of amino acids such as serine, threonine and tyrosine residues on a protein. The most frequently phosphorylated amino acids are serine and threonine. Phosphotyrosine is the infrequent site for phosphorylation. Phosphorylation is done by a group of enzymes known as **kinases** and dephosphorylation is done by oppositely acting enzymes known as **phosphatases**. The phosphorylation database is available for model organisms and humans. Kinases are specific on their substrate selection and their selection is based on the amino acids surrounding the phosphorylation sites. A consensus sequence has been identified on the phosphorylated proteins surrounding the amino acid that is phosphorylated. This sequence is known as the kinase motif. Based on the kinase motif sequence, the kinases are classified into many groups such as proline directed kinases, acidophilic kinases and basophilic kinases. Kinases themselves are phosphoproteins. They activate a signalling cascade by phosphorylating the downstream proteins. An example is MAP kinase pathway. It has been observed that phosphoproteins follow a power law that some proteins accumulate less phosphorylation sites than others during evolution. Phosphorylation often occurs at low stoichiometry or on low abundant proteins, and thus studies of phosphorylated proteins are difficult. Thanks to MS analysis methods, scientists are able to study the phosphoproteins with less difficulty. Many phosphoprotein enrichment techniques are available now to obtain a phosphoprotein-rich sample for studies. A particular kinase cannot phosphorylate all types of proteins. Instead there seems to be compartmentalization of the kinases. The means of studying protein kinases and their substrate is known as **kinome**. Many proteinkinases are identified as drug targets for cancer therapy.

Challenges in phosphoprotein analysis

1. Low stoichiometry of phosphoproteins.
2. Enrichment is needed before phosphoproteins analysis.
3. Assigning functions to the identified phosphoproteins is difficult.
4. Identification of substrate for the phosphoproteins.

15.3 Phosphoprotein enrichment methods

Phosphoproteins can be enriched using the following methods: immunoprecipitation with phosphospecific antibodies, immobilized metal affinity chromatography (IMAC), strong cation exchange (SCX) chromatography, and titanium dioxide chromatographic techniques.

15.3.1 SDS-PAGE separation of phosphoproteins

SDS-PAGE is the most widely used technique for protein separation. In SDS-PAGE all the proteins have net negative charge due to the attachment of SDS to the proteins' molecules. However, there is a difference in SDS attachment between phosphorylated and unphosphorylated proteins. Owing to the negative charge of the phosphoproteins, the amount of SDS binding to the phosphorylated proteins is less than the amount binding to the unphosphorylated proteins. Therefore, the phosphorylated proteins move slowly and the difference between phosphorylated and unphosphorylated proteins is ~80Da. This can be detected and used for mass spectrometry analysis.

15.3.2 Immunoprecipitation

Phosphoproteins can be detected on SDS-PAGE gel or immunoprecipitation using anti-phosphoserine or anti-phosphotyrosine antibodies. Phosphoproteins' detection can also be done using a fluorescent affinity tag (FAT). Fluorescent dyes were specifically attached to the phosphorylation sites by beta-elimination and Michael reactions. Therefore, low abundant phosphoproteins are detected using fluorescent lights. Due to the presence of phosphoproteins with other abundant unphosphorylated proteins, mass spectrometric analysis is difficult as the mass spectrum from other unmodified proteins will over-shadow the actual phosphoprotein signal.

15.3.3 Immobilized metal affinity chromatography

Immobilized metal affinity chromatography (IMAC) is the most widely used method to purify the proteins according to their affinity to specific metal ions, which was first introduced by Porath (1989). This involves the use of phosphate affinity metals which are chelated on resin or beads and packed in a column. The metal is attached to the resin through a linker and the most commonly used linker molecules are nitrilotriacetic acid and iminodiacetic acid. Metals such as Fe^{2+} or Ga^{3+} are chelated on a matrix which is made of silica or agarose to hold the positively charged ions, to which the negatively charged phosphate ions interact. The protein is incubated with the matrix at room temperature and washed with the mixture of methanol and water. Bound peptides are released with excess amount of sodium phosphate or ammonium phosphate.

15.3.4 Strong cation exchange chromatography

In the case of strong cation exchange (SCX) chromatography, separation is done under highly acidic conditions at pH 3.0; in this pH all the carboxylic groups are protonated and only the phosphate group will be negatively charged. Proteins are bound to the negatively charged column and are released in fractions by increasing the salt gradient. Since phosphoproteins are rich in negative charge, they are eluted faster than the unmodified proteins. Non-specific binding of the other unmodified negatively charged proteins to the positively charged IMAC is a common problem. In order to avoid this non-specific binding and to maintain a neutral charge over a wide range of pH, methylation is done to mask the carboxyl side chain of glutamic acid and aspartic acid. Even the enrichment techniques show variability towards the selection of phosphorylated peptides, either multiple phosphorylated or specific phosphorylated proteins. IMAC select multiple phosphorylated peptides on the other hand TiO_2 enrich mono-phosphorylated peptides. Figure 15.1 shows the steps in phosphoprotein analysis.

Published by Woodhead Publishing Limited

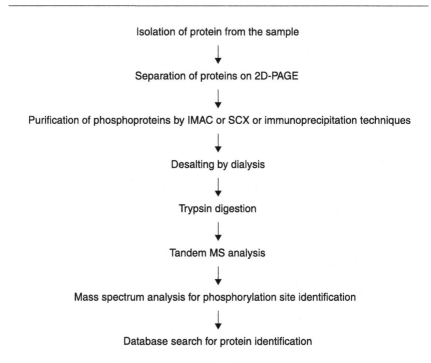

Figure 15.1   Flow chart for steps in phosphoprotein analysis.

15.4 Mass spectrometry for phosphoprotein identification

The ultimate aim of the mass spectrometric analysis of phosphoproteins is to map the phosphorylated site(s). The purified phosphoproteins are digested and supplied to the mass spectrometric analysis. Phosphorylation sites are mapped using **tandem MS** analysis. Mass spectrum of the digested peptide in tandem mass spectrometer generates complex data. From the mass of the subsequent peaks, it is possible to identify the amino acid sequence and also the amino acid which is phosphorylated. *In silico* **digestion** is done for phosphoproteins in the database and the possible phosphoproteins are identified using peptide mass fingerprinting. Mass spectrometer analysis of the phosphoproteins is limited in two important ways (1) mass spectrometry of protein is carried out in positive mode; negatively charged phosphoproteins may not give good spectrum, and (2) the loss of the phosphate group as phosphoric acid from the amino acid residues will not give any information about the site of phosphorylation (Figure 15.2).

Published by Woodhead Publishing Limited

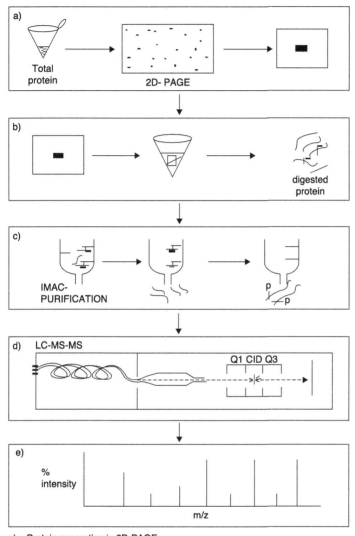

a) Protein separation in 2D-PAGE
b) Protein extraction and digestion
c) Phosphoprotein enrichment
d) Separation of phosphoprotein using liquid chromotography integrated
   with tandem MS
e) Interpretation of MS data to identify the phosphorylation site

Figure 15.2 Phosphoprotein enrichment and analysis. Total protein is separated on 2D-PAGE and the interested spot is excised, the protein on the gel slice is subjected to trypsin digestion, the trypsin-digested proteins are purified using liquid chromatography and subjected to tandem mass analysis. Phosphorylation site is identified from the tandem mass spectrum based on the shift in the position of the amino acid peak.

Published by Woodhead Publishing Limited

Review questions and answers

1. *List the important post-translational modifications.*

- Phosphorylation.
- Glycosylation.
- Acetylation.
- Sulfation.

2. *How does phosphorylation control the enzyme activity?*

Phosphorylation and dephosphorylation are reversible forms of post-translational modification. Some enzymes become active when they are phosphorylated, and otherwise inactive in their dephosphorylated form.

3. *Why is it necessary to enrich the phosphoprotein before mass spectrometric analysis?*

Phosphorylation often occurs at low stoichiometrically or on low abundant proteins. Therefore, studying the phosphorylated proteins is difficult. Many phosphoprotein enrichment techniques are available to obtain rich samples for further studies.

4. *Phosphoproteomics and glycoproteomics give more reliable data to understand gene function. How?*

Proteins are the real executers of cellular function; they take up their function after their post-translational modifications like glycosylation and phosphorylation. Therefore, phosphoproteomics and glycoproteomics give more reliable information than genomics and transcriptomics studies.

Recommended reading

Bentem, S.F., Mentzen, W.I., Fuente, A. and Hirt, H. (2008) 'Towards functional phosphoproteomics by mapping differential phosphorylation events in signaling networks', *Proteomics*, 8: 4453–65.

Gafkin, P.R. and Lampe, P.D. (2006) 'Methodologies for characterizing phosphoproteins by mass spectrometry', *Cell Commun. Adhes.*, 13: 249–62.

Kyong, K.H. and Kristina, H. (2006) 'Selective zirconium dioxide-based enrichment of phosphorylated peptides for mass spectrometric analysis', *Analytical Chemistry*, 78: 1743–9.Larsen, M.R., Thingholm, T.E., Jensen, O.N.,

Roepstorff, P. and Jorgensen T.J. (2005) 'Highly selective enrichment of phosphorylated peptides from peptide mixtures using titanium dioxide microcolumns', *Molecular and Cellular Proteomics*, 4: 873–86.

Mann, M. and Jensen, O.N. (2003) 'Proteome analysis of post-translational modification', *Nature Biotechnology*, 21: 255–61.

Mayya, V. and Han, D.K. (2009) 'Phosphoproteomics by mass spectrometry: insights, implications, applications, and limitations', *Expert Review in Proteomics*, 6: 605–18.

Mumby, M. and Brekken, D. (2005) 'Phosphoproteomics: new insights into cellular signaling', *Genome Biology*, 6: 230.

Pinkse, M.W.H., et al. (2004) 'Selective isolation at the femtomole level of phosphopeptides from proteolytic digests using 2D-NanoLC-ESI-MS/MS and titanium oxide precolumns', *Analytical Chemistry*, 76(14): 3935–43.

Posewitz, M.C. and Tempst, P. (1999) 'Immobilised gallium (III) affinity chromatography of phosphopeptides', *Analytical Chemistry*, 71: 2883–92.

Schmelzle, K. and Forest, M. (2006) 'White phosphoproteomics approaches to elucidate cellular signaling networks', *Current Opinion in Biotechnology*, 17: 406–14.

Woodi, M., Mondal, A.K., Balaram, P. and Krishaswamy, P.R. (2009) 'Analysis of protein post-translational modifications by mass spectrometry: with special reference to haemoglobin', *Indian Journal of Clinical Biochemistry*, 24: 23–9.

16

Glycoproteomics

Abstract: Glycosylation of proteins is an important post-translational modification. The glycan is attached to the protein after synthesis by glycosyl transferases. It is an irreversible form of post-translational modification. In this chapter, different types of glycosylation, glycoprotein enrichment techniques are discussed in detail. Mass spectrometric characterization of glycoprotein is also discussed.

Key words: glycan, glycoproteins, glycosylation, N-linked glycoprotein, O-linked glycoprotein.

Key concepts

- Glycoproteins are conjugated proteins to which sugar molecules are added by a group of enzymes known as glycosyl transferase.
- There are two major types of glycoproteins: O-linked and N-linked, depending up on the amino acid to which the glycan moiety is attached.

16.1 Glycoproteins

Glycoproteins are a common post-translational modification both in prokaryotes and eukaryotes, accounting for more than 50 per cent of all post-translational modifications. Glycoproteins perform many important functions in the cells; their main role is involvement in structural functions in the cell wall or the membrane as receptors. According to the IUPAC definition for glycoproteins, a glycoprotein is a conjugate containing carbohydrate (or **glycan**) covalently linked to a protein. The carbohydrate part of the glycoprotein is known as glycan. The carbohydrate may be monosaccharide (galactose, glucose), oligosaccharides or polysaccharides or their derivatives like sulfo or phosphoro substitutes, aminosugars (N-acetylglucosamine) or acidic sugars (sialic acid). The process of covalent attachment of oligosaccharides is known as **glycosylation**. Since carbohydrates are characterized by the presence of many OH groups, they impart hydrophilicity on proteins. Gylcosylation makes the protein more soluble and helps proper folding of the protein. There is a group of enzymes called glycosyl transferases which add the glycan moiety to the proteins and this usually occurs in the endoplasmic reticulum. Sometimes glycans are attached to proteins spontaneously without the involvement of enzyme. This process is known as glycation. According to the attachment site of glycan moiety to the protein, there are two major groups of glycoproteins called N-linked glycosylation and O-linked glycosylation. In O-linked glycoproteins, the glycan moiety is short and simple and is attached to the OH group of serine or threonine side chain via acyl linkage. In N-linked glycoproteins, the glycan moiety is attached to asparagine via N-acetylglucosamine linkage. The glycan component in N-linked glycoproteins is complex but ubiquitous. There are two types of **N-linked glycans** such as the high mannose type and the acidic type. **O-linked** and N-linked glycoproteins are synthesized through different biosynthetic pathways and their roles are also different.

Glycosylation plays a major role in protein structure and function. Previously glycans were studied after releasing the glycan moiety from protein. But this strategy does not provide any positional information. Thanks to high throughput proteomics techniques like **tandem mass spectrometry**, this helps to characterize both protein and glycan parts of the glycoprotein simultaneously. Purified glycoproteins are used for tandem mass spectrometer analysis. Glycoproteins are also formed at low stoichiometry like phosphoproteins but are irreversible. The other difference is that either monosaccharides or many molecules of carbohydrates are attached to the proteins.

Fractionation is important to enrich the glycoproteins before analysis. The separation of proteins and mass spectrometric analysis of glycoproteins is difficult as the carbohydrates provide a large number of hydroxyl groups which increases the hydrophilicity of the glycoproteins. Glycoprotein analysis is done in two ways; one is analysis of glycopeptides and the other is analysis of glycans only and is called glycomics. About 505 of the mammalian proteins are known to be glycosylated. Many types of life-threatening human diseases like neurodegenerative and immune diseases are associated with improper glycosylation of the proteins. As the glycoproteins are present in low amounts, their detection is difficult. But they are promising diagnostic tools as biomarkers of diseased states and potential drug targets. Glycoproteins account for 50–80 per cent of the mammalian proteins. Missing or aberrant proteins and over-glycosylation of proteins seriously affect their structure and function.

16.1.1 Glycoprotein synthesis

Glycoprotein synthesis occurs in two organelles in sequence such as endoplasmic reticulum and the Golgi apparatus. The carbohydrate core is attached to the protein both co-translationally and post-translationally. The ribosome bearing the mRNA which codes for the proteins attaches to the endoplasmic reticulum. The newly synthesized protein enters the lumen of the endoplasmic reticulum where the addition of the glycan moiety takes place by the action of glycosyl transferase. The glycosylated proteins later enter the Golgi apparatus where trimming of the carbohydrate residues carried out by glycosidase and the final attachment of carbohydrates is done to produce matured glycoproteins which are ready to target their respective destinations.

16.1.2 Functions of glycoproteins

Glycosylation of the proteins is required for proper folding of the proteins. If inhibitors are added to prevent glycosylation, misfolding and protein aggregation occur. The hydrophilic nature of the glycoproteins enhances the solubility of the folding intermediates. Misfolded proteins are cleaved by the protein degradation system of the cell. Sometimes misfolded proteins accumulate and cause serious diseases like Alzheimer's disease or Parkinson's disease. They are also involved in cell signalling and cell adhesion. (This is done by combining both glycan analysis and protein analysis together rather than performing them separately. Studies on glycoproteins are difficult because the glycan moiety is complex; many functional groups and the presence of stereoisomerism make the study difficult.)

16.2 Glycoprotein enrichment methods

Since glycoproteins are present in the cell in an abundant non-glycosylated form, they must be purified before mass spectrometric analysis. Glycoprotein purification can be done by two methods such as lectin affinity chromatography and boronic acid chromatography. The most commonly used lectin is ConcanavalinA (ConA) which has high affinity to unmodified mannosyl or glucosyl groups. It helps to purify the N-linked glycosylated proteins. Another glycoprotein binding protein is wheat germ agglutinin, which binds to N-acetylglucosaminyl and sialic acid. On the other hand, boronic acid method is capable of bonding to *cis-diol* groups of the carbohydrates so that it preferably binds to O-linked oligosaccharides. Glycoprotein enrichment can also be done by using a reverse **glycoblotting** technique.

16.3 Mass spectrometric analysis of glycoproteins

Mass spectrometric analysis of glycoproteins is difficult as they are problematic to volatilize and ionize. The most commonly used mass spectrometry technique is **MALDI-ToF**. Accurate molecular mass determination of the intact glycoprotein before and after removal of the glycan provides information about the type of glycan present in the glycoproteins, but this will not provide information about the site of glycosylation. The glycosylation site can be identified by enzymatic cleavage of the glycans and simultaneous labelling of the asparagines residue by ^{18}O. The glycan can be analyzed by conventional carbohydrate analysis and glycosylation is determined from the MALDI spectrum of the trypsin digested peptides. Another strategy is that the total protein is digested with a non-specific protease and subjected to glycoprotein enrichment. The glycopeptides thus generated are used for MALDI analysis to determine the mass of the glycopeptides. A second set of MALDI is done after removing the glycan moiety. The mass spectrometric data is compared and the site is identified based on the increase mass in the glycosylated peptide. Further structural information about the glycan's moiety in the glycoproteins can be determined by the sequential release of monosaccharide from the glycoproteins using exoglycosidases.

16.4 Importance of glycoproteins in human diseases

Glycoproteins play a part in important cellular functions like embryonic development, cell–cell recognition, cell adhesion, immune functions, and pathogen identification. Many studies show that glycoproteins have a close association with serious human diseases like cancer, rheumatoid arthritis, and immunodeficiency diseases. It has long been recognized that mis-glycosylation plays a major role in diseases and their progression. During oncogenesis expression, the pattern of glycoproteins undergoes

Published by Woodhead Publishing Limited

great changes. There are reports showing increased activity of a particular type of glycosyltransferase V during invasion and metastasis stages of several types of cancers. Glycoprotein analysis in tumour cells helps to find specific glycoproteins that are exclusively expressed in different stages of tumour progress which in turn helps to develop biomarkers for disease diagnosis. Lectin column chromatography of diseased samples helps to determine glycoprotein concentration. Using this technique, many glycoprotein-based biomarkers have been developed for cancer diagnosis. It has also been shown that an aberrant glycosylation pattern is responsible for the development of neurodegenerative diseases like Alzheimer's, Parkinson's, or progressive supranuclear palsy, etc.

Glycoproteomics databases

1. KEGG GLYCAN Structure Database: a collection of experimentally identified glycan structures.
2. KEGG Pathway Maps for Glycans: manually drawn pathways for glycan biosynthesis and degradation.
3. O-glycBase: exclusively devoted to O-linked glycoproteins, this is a non-redundant database with 278 entries.

Review questions and answers

1. *What is a glycoprotein?*

Glycoproteins are bioconjugates synthesized by the addition of carbohydrate moieties to the newly synthesized proteins by glycosyl transferase.

2. *List the different types of glycosylation patterns of proteins.*

- O-linked glycoproteins.
- N-linked glycoproteins.
- Glycosyl phosphotidylinositol proteins.

3. *Name the organelle in which glycosylation occurs.*

The Golgi apparatus.

Published by Woodhead Publishing Limited

Recommended reading

Block, T.M. et al. (2005) 'Use of targeted glycoproteomics to identify serum glycoproteins that correlate with liver cancer in woodchucks and humans', *Proceedings of National Academy of Sciences, USA*, 18: 779–86.

Gamlin, D.P., Scanlan, E.M. and Davis, B.G. (2009) 'Glycoprotein synthesis: an update', *Chemical Review*, 109: 131–63.

Helenius, A. and Aebi, M. (2001) 'Intracellular functions of N-linked glycan', *Science*, 291: 23–64.

Hirabayashi, J., Hashidate, T. and Kasai, K. (2002) 'Glyco-catch method: a lectin affinity technique for glycoproteomics', *Journal of Biomolecular Technique*, 13: 205–18.

Morelle, W. and Michalski, J C. (2005) 'The mass spectrometric analysis of glycoproteins and their glycan structure', *Current Analytical Chemistry*, 1: 29–57.

Pan, S., Chen, R., Aebersold, R. and Brentnal, T.A. (2010) 'Mass spectrometry based glycoproteomics – from a proteomics perspective', *Molecular and Cellular Proteomics*, 123: 13–22.

Taylor, M.T. and Drickamer, K. (2002) *Introduction to Glycobiology*. Oxford: Oxford University Press.

Wuher, M., Catalina, M.I., Deelder, A.M. and Hokke, C.A. (2007) 'Glycoproteomics based on tandem mass spectrometry of glycopeptides', *Journal of Chromatography*, 849: 115–28.

Zhang, H. and Tian, Y. (2010) 'Glycoproteomics and clinical applications', *Proteomics and Clinical Applications*, 4: 124–32.

Web addresses

http://www.cbs.dtu.dk/databases/OGLYCBASE/.
http://www.functionalglycomics.org/glycomics/molecule/jsp/gbpMolecule-home.jsp.
http://www.genome.jp/kegg/glycan/.

Published by Woodhead Publishing Limited

Conclusion

An impressive amount of information has been assembled from the amazing techniques of genome sequencing and this has been deposited in the databases; bioinformatics tools are being developed to extract the useful information for successful exploitation of the genomics and proteomics data. Genome sequencing projects have been underway ever since human genome sequencing was started in 1990. Genomes of many model organisms were sequenced along with the human genome. Later, many more complex genomes were also sequenced in order to exploit their potential. The reason for this notable advancement could be attributed to the developments in sequencing technologies over the years, which resulted in a hundredfold reduction in cost per base. It is expected that the 'next generation sequencing technologies' will reduce the cost of genome sequencing further and help to realize personal genome sequencing. Major challenges comprise developing reliable and sensitive software tools to identify the genes from genome sequencing data, and assigning functions to the genes identified through bioinformatics tools as this often requires manual interference.

The study of proteins is undergoing a radical change from traditional protein chemistry to proteomics, under which many new branches, such as structural proteomics, glycoproteomics, phosphoproteomics, cellular proteomics, clinical proteomics etc., are developing. Coupled with MS instrumentation, protein separation techniques such as 2D-PAGE and LC have provided a detailed analysis of many proteins in a faster and cost effective way. Although genomics and proteomics promise to give solutions to many complex biological problems, much more research has to be carried out to make the techniques and information present in the huge databases useful in various fields, such as healthcare, food and agriculture, environment, industries and forensics. Owing to the development of 'omics', a new branch of biology called 'systems biology' is emerging, which deals with the study of the functions of all genes and their interactions in particular organisms.

Published by Woodhead Publishing Limited

Techniques such as SNP typing, 2D-PAGE, SAGE and microarray have direct applications in disease diagnosis. The high cost, low sensitivity and lack of automation hamper the immediate use of these techniques. For commercial applications, these techniques should be automated, cost must be reduced, and also the entire procedure must be standardized. Major work is needed in the field of post-translational modification of proteins such as glycosylation and phosphorylation, as many human ailments originate due to lack of proper post-translational modifications.

Our challenges for the next five years lie in assigning functions to the unknown genes that are predicted from the human genome sequence. This could be accelerated by comparative genomics; comparing the human genome sequence with other model organism genomes sequenced. This will help to find homologous genes across organisms; if the function of a gene is known in other organisms, a similar function could be assigned to the human counterpart. No doubt genomics and proteomics attracted both academics and biotech industries into the research and development of various fields in the past two decades, as it is expected that such study will provide more detailed understanding of biological problems and help in finding new drug targets for the treatment of disease.

Published by Woodhead Publishing Limited

Glossary

2D-PAGE Separation of proteins in two dimensions based on their isoelectric point and molecular weight in polyacrylamide gel electrophoresis.

Adenine Purine base which forms building blocks of DNA and RNA.

AFLP (amplified fragment length polymorphism) Type of molecular marker used for genetic mapping.

Amplicon PCR amplified DNA fragments.

Anchored primer Mixture of oligonucleotide primers in which (dT) residues are followed by one or two additional bases represented as VN (V = dA or dC or dG and N = dC, dG, dA or dT). This helps the primers to bind to mRNA more efficiently during first strand synthesis.

Antiphosphothreonine Antibody generated against phosphothreonine protein.

Antiphosphotyrosine Antibody generated against phosphotyrosine protein.

Archaebacteria Group of bacteria which live in extreme environments like the deep sea, in a highly alkaline condition, hot springs, etc.

Autonomously replicating sequence (ARS) DNA sequence present in yeast genome responsible for DNA replication.

Autoradiogram Picture of X-ray sheet developed after exposure to radioactive sequencing gel. It has dark bands in lanes which indicate the position of radioactive DNA fragments.

Autoradiography Process of developing X-ray sheet exposed to radioactive material.

BAC (bacterial artificial chromosome) An *E. coli*-based high capacity vector capable of carrying up to 300 kbp foreign DNA segments; it is maintained as single copy plasmid and used for genomic library construction in genome sequencing projects.

BAC library Collection of DNA fragments cloned in BAC vector.

BacMap The database containing genome map of bacteria.

Bacteriophage Viruses which infect bacteria.

Bactig Large DNA sequence obtained by joining or ordering overlapping BAC clones.

Base calling The process by which fluorescence signal is generated in an automated DNA sequencer into nucleotide base. It is done using computer software. One of the earliest bases calling software is Phred.

Base pairing Denonates the hydrogen bonding between nucleotides of complementary DNA strands.

Base peak Highest peak intensity in the mass spectrum.

Capillary electrophoresis A separation technique using a small fused-silica capillary tube (50 μm diameter and one metre length) that allows the use of small volume of sample (10 nL) and high voltage for faster separation.

cDNA Complementary DNA synthesized on an mRNA template by reverse transcription using reverse transcriptase.

Cell The basic structural and functional unit of a living organism. It consists of defined cell membrane, cytoplasm and genetic material.

Cell division Process by which cell is multiplied and it follows defined phases.

CentiMorgan Unit of recombination frequency used to construct the linkage of genetic map to represent the relative order of genes on the chromosome.

Central dogma of life Process which describes the flow of information from DNA to RNA to protein.

Centromere Part of chromosome characterized by heterochromatic nature which is responsible for the separation of chromatids during cell division.

Chemical degradation method of DNA sequencing A method of DNA sequencing in which the order of nucleotides of the DNA fragment is determined from the nested set of DNA fragments that are cleaved by base specific cleavage reactions.

Chromosome Organized structure of genetic material in eukaryotic cells, and present in the nucleus. Each organism maintains a defined number of chromosomes, expressed as diploid or haploid.

CID (collision induced dissociation) A fragmentation process in tandem mass spectrometry in which the selected ions undergo fragmentation in the CID cell.

Clone-by-clone sequencing A genome sequencing method in which genome sequencing is carried out, taking one clone at a time and arranged using a genome map.

Clone fingerprinting A process in which clones sharing overlapping DNA fragments are identified.

Cloning vector Double-stranded DNA molecules with independent origin of replication, selectable marker gene and multiple cloning sites, capable of ligating to a foreign DNA fragment and after placing inside host replicate along with foreign DNA.

Coding region Part of the genome which harbours genes and related regulatory elements for transcription and gene expression control.

Concatemeric DNA Long stretch of DNA formed by ligating many molecules of particular DNA.

Conjugation The transfer of genetic material from one bacteria to another through a conjugation tube.

Contig Large DNA sequence obtained by joining or ordering overlapping clones.

Cosmid *E. coli*-based cloning vector capable of carrying 50 kbp foreign DNA.

Cos site 12 base complementary overhangs present either end of the lambda genome.

Crossing-over A process which occurs during meiosis cell division resulting in exchange of chromatids.

C-value The amount of DNA present in a cell of an organism, epresented in terms of a pictogram.

C-value enigma Conflict between amount of DNA present in an organism and its phenotypic complexity.

Cycle sequencing A sequencing technique in which a nested set of DNA fragments for sequence determination is generated using PCR.

Cytogenetic mapping A physical mapping technique in which the location of a gene or DNA fragments are determined by cytological observation of chromosomes under a microscope.

Cytosine Pyrimidine base which forms the building block of DNA and RNA.

ddNTP (**dideoxy nucleoside triphosphate**) Four types of ddNTPs are ddATP, ddTTP, ddGTP and ddCTP, which are used as a substrate for DNA polymerases.

DDRT-PCR PCR-based gene expression analysis in which differentially expressed genes are identified.

de novo **protein sequencing** A protein sequencing method in which a sequence of amino acids of polypeptide is determined from tandem mass spectrometric data.

Deoxyribose A five-carbon monosaccharide in which C2' does not have a hydroxyl group.

Dephosphorylation Process by which the phosphate group is removed to the proteins by kinases.

Depth of coverage The number of times a particular part of the genome is sequenced.

Depurination The state in which purine bases are selectively removed.

Dideoxy chain termination sequencing method A method of DNA sequencing in which the order of nucleotides of the DNA fragment is determined from the nested set of DNA fragments that are terminated by dideoxy nucleoside triphosphates.

Diploid These cells have two sets of chromosomes and this is denoted by 2n.

DNA (**deoxyribonucleic acid**) Made up of nucleotides which make up the genetic material of many organisms.

DNA polymerase Enzyme responsible for synthesis of a new DNA strand.

DNA repair The process by which damaged DNA or mutated DNA is corrected.

DNA sequencing Determining the order or arrangement of nucleotides in a DNA fragment.

DNase Enzyme which cleaves DNA fragments.

dNTP (**deoxy nucleoside triphosphate**) Four types of dNTPs are dATP, dTTP, dGTP and dCTP which are used as substrate for DNA polymerases to synthesize a new DNA strand.

DOE (Department of Energy) Located in Washington, DC, conducted programme related to energy-related issues of the US, through its radiation safety programme. DOE initiated the Human Genome Sequencing Project and funded the largest portion of the money for the publicly funded HGP.

Draft genome sequence The genome sequence published with gaps and ambiguous bases.

Dye terminator Fluorescent dye attached to ddNTP, which are used in automated DNA sequencing.

Edman degradation A process by which N-terminal amino acid in a polypeptide is determined using phenyl isothiocyanate (PITC).

Edman reagent Phenyl isothiocyanate (PITC) which reacts with N-terminal amino acid of a protein to form modified amino acid.

EI An ionization technique in which a sample is passed perpendicular to the high energy electron beam to obtain the gas phase ions.

Electro-blotting Process of transferring charged molecules such as a protein from gel to the membrane using an electric current.

Electrophorogram Graphical representation of the sequencing picture in intensity vs time lapse.

Electroporation Method of introducing foreign DNA into a host by electric shock.

ELSI Ethical, legal, and social issues related to human genome sequencing and its possible effect on human beings.

Emulsion PCR Type of PCR in which amplification is carried out in an oil–water interface.

End-labelling The process of incorporating radioactive phosphate to either the 3' or 5' end of the DNA strand.

Endocytosis The process by which viruses or other outside particles are taken inside the eukaryotic cells.

Endonucleases Enzymes which are capable of cleaving the phosphodiester bonds into DNA fragments.

Endoproteases A group of enzymes capable of cutting the peptide bonds in a polypeptide chain.

Enhancer DNA sequence present in the vicinity of genes which increases the transcription activity.

ESI (electrospray ionization) Soft ionization technique in which the protein sample is passed through the heated capillary tube and sprayed as droplets. When the droplets evaporate, the ions are released in the gaseous form.

EST (expressed sequence tag) A partial sequence of cDNA fragment, deposited in EST database. Used for expression studies.

Euchromatin Part of a genome which is transcriptionally active.

Eukaryotes Cells with complex cellular structures and defined nucleus surrounded by nucleolar membrane.

Exon A coding part of gene represented in the mRNA.

Exonuclease Enzyme which degrades one nucleotide at a time from the ends of a polynucleotide strand.

Expression profiling Cataloguing of all genes expressed in a cell or tissue.

Extrachromosomal DNA The DNA present in a cell other than main genome of an organism. Example is plasmid DNA present in certain bacterial cells.

FAB (fast atom bombardment) An ionization technique in mass spectrometry in which the analytes are mixed with non-volatile chemicals and bombarded with Xenon gas.

F factor A type of plasmid present in *E. coli*, determines the sex of the *E. coli* cell and promotes conjugation between F+ and F- bacteria.

Finished genome sequence Completed genome sequence where each base is sequenced more than 10 times which has a low error rate and minimum gaps.

FISH (fluorescent *in situ* hybridization) A technique in which separated chromosomes are placed on a microscopic slide and hybridized with a fluorescently labelled probe to locate the position of the DNA sequence on the chromosome.

FlyBase Database containing all genetic, genomic information about *Drosophila melanogaster.*

Forward genetics Studying phenotype and then identifying gene responsible for it.

Fosmid *E. coli*-based cloning vector capable of carrying 80 kbp–100 kbp foreign DNA.

Functional genomics Branch of genomics which deals with finding functions of genes from DNA or mRNA or protein sequences.

Published by Woodhead Publishing Limited

Gene Unit of heredity that controls a particular trait of an organism. A fragment of DNA that codes for a particular protein.

Gene expression A process by which DNA is transcribed to RNA, which is subsequently translated to a protein.

Gene knock-out A genetic manipulation process by which a gene is deleted from the genome of an organism to study its role.

Genetic manipulation A process by which genome of an organism is artificially modified.

Genetic map Description of genome in terms of genes or molecular markers and their relative order and distance. The unit of genetic map distance is a centiMorgan.

Genetics Science dealing with study of heredity and variations in living organisms.

Genome The entire complement of DNA present in a haploid cell.

Genome annotation Process of identifying genes, regulatory sequences and other features of a genome from its sequence.

Genome assembly Process of assembling the DNA fragments after sequencing.

Genome map Description of a genome in terms of order and relative location of landmarks such as gene, molecular markers, restriction sites, STS markers, etc.

Genome model An organism which is used for genetic studies such as genome sequencing.

Genome sequencing The process of determining the order of nucleotides that constitute the genome of an organism.

Genomics A branch of genetics which deals with mapping, sequencing and annotation of genomes.

Glycan A group of polysaccharides or carbohydrates which are added to the proteins to form glycoproteins.

Glycoblotting Process by which glycoproteins are transferred to the membrane after separating them on polyacrylamide gel.

Glycoprotein Conjugated protein in which a glycan moiety is covalently attached.

Glycoproteomics Branch of proteomics which deals with studies on glycoprotein structure and function.

Published by Woodhead Publishing Limited

Glycosylation Process by which carbohydrate or glycan is added to the newly synthesized proteins.

Guanine A purine base which forms the building block of DNA and RNA.

Haploid Number of chromosomes present in the gamete or sex cells of an organism.

Haplotype DNA sequences that are shared by multiple individuals in a population.

Heterochromatin Condensed part of a genome that does not take part in transcriptional activity.

High capacity vector Cloning vectors capable of carrying large DNA fragments, i.e. more than 100 kbp.

Histones Group of proteins found in the nucleosomes of higher eukaryotes which are responsible for the packaging of chromosomes.

Homologous genes Genes present in two or more organisms which are evolved before speciation.

HUGO Human Genome Organization, founded on 30 April 1988 to coordinate the international effort to sequence the human genome.

Human genome The total amount of DNA present in a haploid human cell.

Human Genome Project (HGP) Large biological project to determine the order of nucleotides in the human genome.

Immunoprecipitation Process by which protein (antigen) is precipitated with antibody.

Intron Non-coding region of the gene which is removed during splicing.

in silico **digestion** Virtual digestion of database protein sequence to generate peptides.

in vitro **packaging** Assembling of matured phage particles from prehead, tail and concetemeric DNA.

Isoelectric focusing Method of separating proteins based on their isoelectric point.

Isoelectric point pH at which a particular protein carries no net charge.

Karyotype The pictorial representation of chromosomes of a eukaryotic organism according to their size.

Kinases Group of enzymes catalyzing the addition of phosphate group to the proteins.

Published by Woodhead Publishing Limited

Kinome The entire complement of phosphoproteins expressed in a cell.

Lambda phage Type of *E. coli* bacteriophage.

Lectin affinity chromatography Type of affinity chromatographic techniques in which lectin (carbohydrate binding proteins) are used to separate the glycoproteins.

MALDI (matrix assisted laser desorption and ionization) A soft ionization method in which protein to be analyzed is co-crystallized with small organic solvents and dried. Laser beam is applied to the co-crystal to ionize as well as to convert the sample to gas phase.

Mass range Limit of the mass spectrometer to detect lowest and highest mass to charge ratio of ions.

Mass resolution Ability of the mass spectrometer to differentiate two fragments with close mass to charge ratio.

Mass spectrometry An analytical technique in which gas phase charged molecules are separated and detected according to their mass to charge ratio.

Metabolomics Branch of science deals with small molecular weight metabolic products and their functions in a cell.

Microarray A solid support with thousands of probe sequences (cDNA) immobilized in rows and columns used for nucleic acid hybridization analysis.

Microelectrophoresis An electrophoretic technique in which a micro-electrophoretic device is used for DNA fragment separation.

Model organism An organism which is being studied extensively and the information derived from it will be of great use for human beings.

Molecular marker A polymorphic DNA sequence which can be mapped on genomic DNA using molecular techniques such as Southern blotting, PCR, etc.

Monoisotopic mass Mass calculated using only most abundant isotopic elements of molecule.

mRNA(messenger RNA) RNA synthesized on template DNA by RNA polymerase which has information to code for polypeptide.

Multiple cloning sites DNA fragment in which a number of restriction enzyme recognition sequences are closely located on a cloning vector.

Published by Woodhead Publishing Limited

NIH (National Institute of Health) Located in Maryland, USA, was established in 1887 within the US Public Health Service. It played a significant role in publicly funded HGP.

N-linked glycan A type of glycoprotein conjugate in which the glycan moiety is attached to the amino group of asparagine residue of a protein.

Non-coding region Part of the genome which harbours repetitive DNA sequences and does not take part in transcription activity.

Northern blotting Hybridization-based gene expression analysis technique in which RNA is transferred to a nylon membrane using capillary blotting.

Nucleoid Nucleus-like structure present in prokaryotes which stores the genetic information.

Nucleoside Biomolecule made of ribose sugar and base nitrogenous linked through glycosidic bond.

Nucleosome Basic building block of chromosome made up of DNA and histone proteins.

Nucleotide Basic building block of the DNA strand, made up of five carbon sugar, nitrogenous base and phosphate group.

Nucleus Cellular organelle in eukaryotes with well-developed membrane which stores the genetic information.

Oligo(dT) affinity chromatography Type of chromatography in which oligo(dT) is covalently attached to a solid support and used to purify the mRNA.

Oligonucleotide Synthesized by phosphodiester linkage of 10 to 15 nucleotides.

Oligosaccharide Formed by glycoside linkage of 3 to 10 monosaccharides

O-linked glycan A type of glycoprotein conjugate in which the glycan moiety is attached to the hydroxyl group of serine or threonine residues of a protein.

Operator Upstream DNA sequence which regulate on and off state of a gene.

Operon Arrangement of functionally related genes under a single promoter which are transcribed and translated together.

Optical mapping Type of restriction mapping method in which restriction sites on large DNA fragments are detected under the microscope.

ORF (open reading frame) Coding part of the gene, it begins with start codon and ends with stop codon.

Origin of replication DNA sequence on plasmid or genome from which DNA replication is initiated. It is a *cis* acting DNA sequence.

Orthologous genes Genes present in two or more organisms which are evolved from a common ancestor.

P element A transposon found in *Drosophila*.

P1 phage Type of *E. coli* bacteriophage.

Paralogous genes Genes present in an organism which is evolved from a common ancestor after speciation.

PCR (polymerase chain reaction) *In vitro* amplification of DNA fragments using DNA polymerase.

Peak list List of masses generated from mass spectrum of peptides of a protein.

Pedigree analysis Study of inheritance pattern of genes in related individuals.

PFGE (pulse field gel electrophoresis) A type of electrophoresis technique by which a large number of DNA fragments are separated.

Phenomics Study of total phenotypes of an organism.

Phosphatase Enzyme catalyses the removal of phosphate group from the DNA fragments.

Phosphoproteins Proteins in which one or more amino acids are phosphorylated.

Phosphorylation Process by which phosphate group is added to the proteins by kinases.

Phosphoserine A phosphorylated protein in which the phosphate group is added to the serine residue of a polypeptide.

Phosphothreonine A phosphorylated protein in which the phosphate group is added to the threonine residue of a polypeptide.

Phosphotyrosine A phosphorylated protein in which the phosphate group is added to the tyrosine residue of a polypeptide.

Plasmid Double-stranded self-replicating DNA present in certain bacterial cells as extrachromosomal DNA.

PMF (peptide mass fingerprinting) A protein identification technique in which the accurate masses of the peptides generated from an unknown protein are compared with the peptide peak list generated from the database.

Polynucleotide Biopolymer formed by the covalent linkage of nucleotides.

Polypeptide Biopolymer formed by the covalent linkage of amino acids.

Polyploid Cells having more than two sets of chromosomes.

Polysaccharide They are formed by glycoside linkage of more than 10 monosaccharides.

Polytene chromosome A special type of chromosome found in salivary glands of *Drosophila* which are duplicated many times without being separated.

Post-translational modification (PTMP) The process by which the newly synthesized proteins undergo different modifications such as phosphorylation, glycosylation, sulfation.

Primer Short oligonucleotide that helps to initiate DNA synthesis by DNA polymermase.

Prokaryotes Cells with simple cellular structures and a primitive form of nucleus not surrounded by a nucleolar membrane.

Promoter A DNA sequence present in the upstream side of a gene which is needed for transcription to start.

Protein Polymer of amino acids which executes the various structural and physiological functions of living organisms.

Protein folding Process by which proteins undergo the ordered formation of secondary structures such as alpha helix and beta sheets.

Proteolytic cleavage Process by which proteins are cleaved at particular sites by specific proteases.

Pyrosequencing Sequencing method in which the base sequence of a DNA fragment is identified during synthesis of complementary sequence.

Quadrupole analyzer Type of mass analyzer present in the mass spectrometer that separates ions based on their stability to applied electric field along the rods.

Quantitative proteomics A branch of proteomics dealing with quantification of all proteins expressed in a cell or tissue.

Radiation hybrid A hybrid cell produced by fusing a rodent cell with an irradiated human cell, used in physical mapping technique.

Radioactive labelling A method in which a radioactive isotope of phosphate is added to a DNA fragment by enzymes such as kinase.

Random primer A short oligonucleotide whose sequence is selected randomly.

RAPD (randomly amplified polymorphic DNA) Type of molecular marker characterized by PCR technique using random single primer.

RDA (representational difference analysis) Gene expression analysis method in which unique cDNA are subtracted and amplified using PCR.

Recombination frequency Measure of genetic linkage between genes, calculated from the progenies of the breeding experiment. It is the number of recombinants observed to the total number of progenies.

Repeat sequence A kind of DNA sequence in which particular bases are repeated many times in tandem.

Replication Process by which the genome of an organism is duplicated before every cell division by the action of DNA polymerases.

Restriction mapping A method by which the relative location of restriction sites on a DNA fragment are determined by restriction digestion followed by agarose gel electrophoresis.

Retrovirus The group of viruses harbouring RNA as their genetic material.

Reverse genetics Identifying a phenotype or function from the gene sequence.

RFLP (restriction fragment length polymorphism) Type of molecular weight marker generated by Southern hybridization.

Ribose Five carbon monosaccharide present in ribonucleic acid.

RNA (ribonucleic acid) Made of ribose sugar, nitrogenous bases and phosphate group. It is a genetic material in some viruses.

SAGE (serial analysis of gene expression) Sequencing-based high throughput gene expression analysis method.

SDS-PAGE (sodium dodecyl sulphate–polyacrylamide gel electrophoresis) A separation technique for proteins in which proteins are separated based on their molecular weight.

Segmented duplication Presence of identical or near-identical DNA sequences on chromosomes which are created through chromosomal duplications.

Selectable marker gene A gene code for a product which is capable of conferring resistance on the host in the presence of a selection agent.

Published by Woodhead Publishing Limited

Sequencing gel A high percentage polyacrylamide gel which is capable of separating DNA fragments with one nucleotide difference.

SH (subtractive hybridization) A hybridization-based gene expression analysis technique to isolate tissue specific genes.

Southern blotting Molecular technique in which a specific DNA fragment is detected among other DNA fragments by nucleic acid hybridization on a solid support using a labelled probe.

Split gene Eukaryotic genes which are divided into coding and non-coding regions.

STS (sequence tagged site) A type of molecular marker used for physical mapping techniques developed from known DNA sequence.

STS content mapping A physical mapping technique in which STS markers are ordered on large DNA clones such as a BAC clone or a YAC clone using molecular techniques like PCR.

Synteny Conservation of gene order in genomes of different organisms.

Synthetic carrier ampholytes Small molecules bearing both positive and negative charges.

Tandem mass spectrometry Type of mass spectrometry in which ions are separated in two stages; in the first stage, only one ion of specific mass to charge is selected, and in the CID cell the selected ion is fragmented and the fragmented ions are further separated in the second analyzer. It is useful in protein sequence determination.

Telomere A part of the eukaryotic chromosomes located at ends which are characterized by the presence of highly repetitive DNA sequences.

Terminator A DNA sequence present in the downstream of a gene which helps to terminate the transcription of a gene.

Thymine A pyrimidine base which forms the building block of DNA.

ToF analyzer (time of flight) Type of mass analyzer present in mass spectrometer, it separates the ions based on flight time from one end of the analyzer to the other end.

Transcription Process by which the RNA is synthesized from the DNA template by RNA polymerase.

Transcriptomics Study of total number of transcripts that are expressed in a particular cell or tissue.

Transduction Method of introducing foreign DNA into bacteria using bacteriophage.

Transformation Method of introducing foreign DNA into bacteria by a conjugation tube formed between two bacterial cells.

Translation Synthesis of polypeptide mRNA as template.

tRNA (transfer RNA) Group of RNA responsible for bringing charged amino acids to the ribosome assembly during translation.

Ubiquitination A post-translational modification of small proteins called ubiquitins.

WormBase Database containing all genetic information about *C.elegans*.

YAC (Yeast Artificial Chromosome) Yeast-based cloning vector capable of carrying a large foreign DNA fragment.

Published by Woodhead Publishing Limited

Index

Published by Woodhead Publishing Limited

Printed in the United States
By Bookmasters